# Updates on Endoplasmic Reticulum

*Edited by Gaia Favero*

Published in London, United Kingdom

Updates on Endoplasmic Reticulum
http://dx.doi.org/10.5772/intechopen.100952
Edited by Gaia Favero

Contributors
Bibu John Kariyil, Priyanka Menon Kunnel, Sikander Ali, Maria Najeeb, Milan Fiala, Mary Dover, Michael Kishek, Miranda Eddins, Naneeta Desar, Ketema Paul, María Isabel Hernández-Alvarez, Raúl Ventura, Bharathidevi Subramaniam Rajesh, Ramya Ravi, Eduard Korkotian, Liliya Kushnireva, Vishwa Jyoti Baruah, Manny Saluja, Bhaswati Sarmah, Elizabeth H. Mahood, Gaia Favero, Francesca Bonomini, Marzia Gianò, Rita Rezzani



First published in London, United Kingdom, 2023 by IntechOpen
IntechOpen is the global imprint of INTECHOPEN LIMITED, registered in England and Wales, registration number: 11086078, 5 Princes Gate Court, London, SW7 2QJ, United Kingdom
Printed in Croatia

British Library Cataloguing-in-Publication Data
A catalogue record for this book is available from the British Library

Additional hard and PDF copies can be obtained from orders@intechopen.com

Updates on Endoplasmic Reticulum
Edited by Gaia Favero
p. cm.

This title is part of the Biochemistry Book Series, Volume 46
Topic: Cell and Molecular Biology
Series Editor: Miroslav Blumenberg
Topic Editor: Rosa María Martínez-Espinosa

Print ISBN 978-1-80356-227-8
Online ISBN 978-1-80356-228-5
eBook (PDF) ISBN 978-1-80356-229-2
ISSN 2632-0983

IntechOpen Book Series

# Biochemistry

Volume 46

## Aims and Scope of the Series

Biochemistry, the study of chemical transformations occurring within living organisms, impacts all of the life sciences, from molecular crystallography and genetics, to ecology, medicine and population biology. Biochemistry studies macromolecules - proteins, nucleic acids, carbohydrates and lipids –their building blocks, structures, functions and interactions. Much of biochemistry is devoted to enzymes, proteins that catalyze chemical reactions, enzyme structures, mechanisms of action and their roles within cells. Biochemistry also studies small signaling molecules, coenzymes, inhibitors, vitamins and hormones, which play roles in the life process. Biochemical experimentation, besides coopting the methods of classical chemistry, e.g., chromatography, adopted new techniques, e.g., X-ray diffraction, electron microscopy, NMR, radioisotopes, and developed sophisticated microbial genetic tools, e.g., auxotroph mutants and their revertants, fermentation, etc. More recently, biochemistry embraced the 'big data' omics systems. Initial biochemical studies have been exclusively analytic: dissecting, purifying and examining individual components of a biological system; in exemplary words of Efraim Racker, (1913 –1991) "Don't waste clean thinking on dirty enzymes." Today, however, biochemistry is becoming more agglomerative and comprehensive, setting out to integrate and describe fully a particular biological system. The 'big data' metabolomics can define the complement of small molecules, e.g., in a soil or biofilm sample; proteomics can distinguish all the proteins comprising e.g., serum; metagenomics can identify all the genes in a complex environment e.g., the bovine rumen.

This Biochemistry Series will address both the current research on biomolecules, and the emerging trends with great promise.

# Meet the Series Editor

Miroslav Blumenberg, Ph.D., was born in Subotica and received his BSc in Belgrade, Yugoslavia. He completed his Ph.D. at MIT in Organic Chemistry; he followed up his Ph.D. with two postdoctoral study periods at Stanford University. Since 1983, he has been a faculty member of the RO Perelman Department of Dermatology, NYU School of Medicine, where he is codirector of a training grant in cutaneous biology. Dr. Blumenberg's research is focused on the epidermis, expression of keratin genes, transcription profiling, keratinocyte differentiation, inflammatory diseases and cancers, and most recently the effects of the microbiome on the skin. He has published more than 100 peer-reviewed research articles and graduated numerous Ph.D. and postdoctoral students.

# Meet the Volume Editor

Dr. Gaia Favero is a researcher for the scientific-disciplinary sector BIO/16 Human Anatomy at the Anatomy and Physiopathology Division, Department of Clinical and Experimental Sciences, University of Brescia, Italy. Her research programs focus on aging-related morphological dysfunctions as the prelude to various pathophysiological processes, and the central hypothesis that natural antioxidants and melatonin may act as molecular "switches" that modulate cells and tissues by suppressing oxidative stress and inflammatory signaling pathways. These research approaches represent a powerful tool to develop innovative preventive and/or therapeutic strategies and to identify novel prognostic biomarkers for several diseases. Dr. Favero is the author of seventy-five manuscripts published in various peer-reviewed international journals.

# Contents

# Preface

The endoplasmic reticulum is a complex and multifunctional organelle present in the cytosol of eukaryotic cells. This organelle is not only essential for calcium homeostasis but also for lipid biosynthesis and protein assembly, folding, and post-translational modification. The endoplasmic reticulum contains domains that control the interaction with other organelles, such as the Golgi apparatus, endosomes, mitochondria, lipid droplets, and peroxisomes. It has been established that energy/nutrient depletion, calcium flux injury, or oxidative stress disrupt endoplasmic reticulum homeostasis and even induce accumulation of misfolded/unfolded proteins leading to endoplasmic reticulum stress. Under conditions of endoplasmic reticulum stress, an adaptive mechanism of coordinated signaling pathways, defined as the unfolded protein response, is activated to return the endoplasmic reticulum to its healthy functioning state.

During aging, there is a noted decrease of the protective adaptive response of the unfolded protein response and an increase of the pro-apoptotic pathway together with endoplasmic reticulum ultrastructural injury. Controlling endoplasmic reticulum stress response, maintaining the appropriate endoplasmic reticulum ultrastructure and homeostasis, and retaining organelles' interplay are crucial aspects of global cellular health.

This book presents a comprehensive overview of the endoplasmic reticulum, including endoplasmic reticulum "ultrastructural anatomy," organelles' interplay, endoplasmic reticulum stress and its implication in plant and human health and disease conditions, and more. The information contained herein can help identify and develop novel therapeutic approaches to endoplasmic reticulum response, which may lead to early detection of correlated stressful and/or disease states.

Dr. Favero Gaia would like to thank all the authors who contributed to the success of this book and the IntechOpen team for their valuable and constant support.

**Gaia Favero**
Anatomy and Physiopathology Division,
Department of Clinical and Experimental Sciences,
University of Brescia,
Brescia, Italy

Section 1

# Endoplasmic Reticulum in Healthy State

Chapter 1

# Navigating the Endoplasmic Reticulum: New Insights and Emerging Concepts

*Sikander Ali and Maria Najeeb*


## Abstract

Endoplasmic reticulum (ER) is a membrane bound organelle adjacent to the nucleus in eukaryotic cells. It exists in the form of membranous sacs called "cisternae". It was first discovered by Emilio Veratti in 1902 and later named as 'Endoplasmic Reticulum' in 1953 after visualization through electron microscopy. There are two types of endoplasmic reticulum based on the presence of ribosomes i.e., 'rough' ER and 'smooth' ER. Rough ER is the site for protein synthesis and modification by glycosylation. While the smooth ER is involved in the metabolism of lipids and carbohydrates. Recently, it has been classified on the basis of membrane structure rather than appearance. It physically interconnects with the mitochondria and these sites are named as mitochondria-associated membranes (MAMs) that are crucial for $Ca^{+2}$ homeostasis. Various mechanisms of ER signaling play vital role in physiology and the onset of disease. A thorough understanding of these mechanisms and their role in physiology and pathophysiology can be applied to develop new ER-targeted therapies.


**Keywords:** endoplasmic reticulum (ER), mitochondria-associated membranes (MAMs), $Ca^{+2}$ homeostasis, ER-targeted therapies, perturbic ER functions

## 1. Introduction

Endoplasmic reticulum (ER) is a membrane bound subcellular organelle that appears as a network of tubules in the cytoplasm [1]. It was first observed in chicken fibroblast-like cells with the help of electron microscope. After 10 years, it was named by Porter in 1954 [2]. There are two types of ER based upon the presence or absence of ribosomes on its surface i.e., smooth endoplasmic reticulum SER and rough endoplasmic reticulum RER. Both of these could occur either interconnected or separately in compartments [3]. ER is a crucial site for various functions, including protein synthesis, storage and regulation of $Ca^{2+}$ and lipid and glucose metabolism [3]. These range of functions indicate that ER plays important role in regulating metabolism and cell programming.

ER plays important role in protein synthesis and modification. After synthesis the proteins are translocated across the membrane of ER. Ribosomes attached to the RER membrane are involved in the synthesis of protein [4]. In addition, ER is also crucial

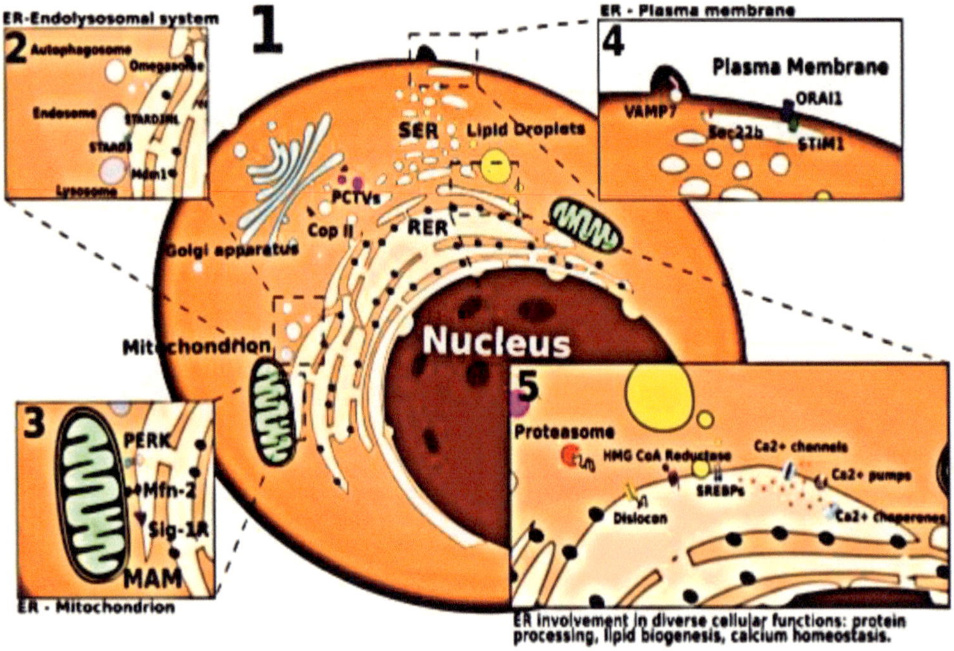


**Figure 1.**
*ER molecular machines and contact sites with other organelles. The ER forms multiple membrane contact sites with other organelles, including the endosomes and lysosomes (through STARD3, STARD3NL, Mdm1; panel 2), the mitochondria (through Mfn-2, Sig-1R, PERK; panel 3), and the PM (through ORAI1, STIM1, Sec22b, VAMP7; panel 4) with various functional implications.*

for $Ca^{2+}$ signaling and homeostasis. A storage and transport system for $Ca^{2+}$ comprising of $Ca^{2+}$ channels, ATPase pumps, sensors and storage proteins. This chapter highlights a new aspect for ER classification, its functions, ER phagy and role of ER signaling in health and disease [5].

## 2. Structure

There are two types of ER based upon the presence or absence of ribosomes on its surface i.e., smooth endoplasmic reticulum SER and rough endoplasmic reticulum RER [2]. Both of these could occur either interconnected or separately in compartments. Recently, a new classification has been introduced based on its membrane structure instead of appearance. According to this, ER carry a nuclear envelope, cisternae and three way connected tubules. The ER is interconnected with many organelles in the cell and it is connected with mitochondria through specific sites, called mitochondria-associated membranes (MAMs), that are critical for maintaining $Ca^{2+}$ homeostasis [6]. Its interaction with the plasma membrane is regulated by stromal interaction molecule 1 a protein-like structure and Calcium channel protein 1 [7]. Moreover, SEC22b a vesicle-trafficking also involved in the maintenance of this interaction [8]. ER interaction with endosomes is stabilized by lipid transfer protein 3, the protein involved in cholesterol maintenance in endosomes [9]. ER also play role in autophagy by interacting with endolysosomal system. These intracellular interactions are crucial for the functionality of the cell (**Figure 1**) [10].

## 3. Functions

ER is a crucial site for various functions, including protein synthesis, storage and regulation of $Ca^{2+}$ and lipid and glucose metabolism. These range of functions indicate that ER plays important role in regulating metabolism and cell programming [3].

### 3.1 Protein modification

Secretory and transmembrane proteins are synthesized in ER. Their folding, maturation, quality control and degradation also occur in this organelle. As a result, only properly folded proteins are transported to their destination [4]. Approximately, 30% proteins are cotranslationally delivered to the ER, where chaperones are involved in their folding, packaging and post-translational modification [11]. Protein modification processes includes signal sequence cleavage, formation and breakage of disulfide bonds and lipid conjugation. Misfolded proteins are damaging to normal cellular functions and are tightly monitored. Protein misfolding is a regular process but aggravate during adverse conditions. In ER several regulatory systems ensure correction of misfolded proteins. Terminally misfolded secretory proteins are removed by ER associated degradation (ERAD) process [12]. Initially, proteins are encountered by an ER resident luminal and transmembrane protein machinery, then translocated into cytosol via channel called dislocon [13].

### 3.2 Lipid synthesis

ER is also crucial for synthesis of membrane constituents, lipid droplets fat accumulation as energy reservoir. Lipid synthesis takes place at membrane interfaces and organelle interaction sites. The ER membrane architecture dynamically altered according to cellular lipid concentration [14]. In order to maintain the cholesterol homeostasis in the body, ER carry a family of protein sensors. It also consists of various enzymes involved in synthesis of sterols and phospholipids (**Figure 1**) [15].

### 3.3 ER export

The proteins and lipids synthesized in the ER are delivered to their destined locations through secretory pathways. The export process is tightly controlled to maintain a steady anabolic flux because anomalies in secretion could lead to detrimental consequences to ER structure and functions [16]. Synthesis of ER COPII transport vesicles is crucial to this export process. Apart from COPII mediated transport several other mechanisms have been studied such as nonvesicular. For example, large lipoprotein cargo is transported out in another vesicle or stored in lipid droplets (**Figure 1**) [17].

### 3.4 $Ca^{2+}$ homeostasis

$Ca^{2+}$is a metallic ion plays key role as a secondary messenger in various intracellular and extracellular signaling events such as gene expression, translation, protein trafficking, and regulation of other cellular functions [6]. ER is the main reservoir of $Ca^2$+ and important for its regulation. Myriad of cellular functions are regulated by $Ca^{2+}$-dependent way in order to maintain the calcium level of entire cell. As a result, tight regulation of both ER and cytoplasmic $Ca^{2+}$ concentration is essential to maintain enhanced intraluminal Ca2+ concentration redox potential as compared to the cytoplasm. A variety of mechanism employed by ER to maintain the concentration of $Ca^{2+}$ inside and outside the membrane [5]. (a) ER membrane ATP-dependent $Ca^{2+}$ pumps for cytosol-to-lumen transport; (b) ER luminal $Ca^{2+}$-binding chaperones for sequestering free $Ca^{2+}$ and (c) ER membrane channels for the regulated release of $Ca^{2+}$ into the cytosol. These mechanisms are supported by a controlled interaction between the ER and other organelles, i.e., PM and the mitochondria.

## 4. Perturbing ER functions

Disturbance in ER function results in condition commonly known as 'ER stress'. In order to overcome ER stress and reestablish the homeostasis several adaptation mechanisms are activated inside the cell [18].

### 4.1 Intrinsic ER perturbations

In certain disease conditions such as cancer, diabetes and neurodegenerative diseases some cellular mechanisms lead to ER stress [19]. In case of cancer, rapid and uncontrolled cellular growth required high protein production that impact the ER system [20]. More specifically, in melanoma that has highest mutation rate the number of mutated proteins is increased that results in ER stress. In chronic myeloid leukemia, an oncoprotein is activated that promotes cell proliferation and disturbs $Ca^{2+}$-dependent apoptotic response [21].

Many neurodegenerative diseases also disturb the ER homeostasis and lead to ER stress. For instance, motor neuron death is the consequence of mutations in the vesicle-associated membrane protein-associated protein B located in ER. It is mediated by the fluctuation of ER stress signaling [22, 23]. On the contrary, pancreatic beta cells involved in insulin production carry a complex and developed ER to control insulin production and use in response to high blood sugar level. Type 1 diabetes is associated with mutation induced ER stress, in this condition beta cells undergo apoptosis and insulin level reduced [24, 25]. Insulin mutation-related ER stress have also been observed in neonatal diabetes [26, 27].

### 4.2 Extrinsic ER perturbations

#### *4.2.1 Microenvironmental stress*

Microenvironmental ER stress occurs in tumorigenic cells. These cells rapidly proliferate that leads to deprivation of nutrients and oxygen in the microenvironment, resulting in local stress accompanied with hypoxia, starvation and acidosis, consequently ER stress, perturbation of protein and lipid biogenesis [28]. Nutrient scarcity, most importantly glucose distress facilitates ER stress by perturbing glycosylation.

#### *4.2.2 Exposure to ER stressors*

ER stressors are small molecules that stimulate ER stress mediated by a number of mechanisms [29]. These include molecules such as tunicamycin [30], or 2-deoxyglucose target the N-linked glycosylation of proteins. On the other hand, dithiothreitol prevents protein disulfide bond formation [31]. While Brefeldin A inhibits the ER to-Golgi trafficking, resulting in rapid and reversible inhibition of protein secretion [32].

#### *4.2.3 Exposure to enhancers of ER homeostasis*

Some molecules have been reported to stimulate and increase ER stress, such as peptides and proteostasis regulators. A most commonly used 4-phenylbutyric acid (4-PBA) prevents the aggregation of misfolded proteins in the ER [33]. In islet cells, to reduced ER stress a bile acid called Tauroursodeoxycholic acid (TUDCA) is present [34]. FDA has approved TUDCA as a drug for patients diagnosed with primary biliary cirrhosis [35]. The precise mechanism of action of these proteostasis regulators is still unknown.

#### *4.2.4 Temperature*

Mammals normal body temperature is 36–37°C necessary for viability and normal bodily functions. Fluctuations in normal body temperature could perturb cellular homeostasis consequently protein denaturation and aggregation [36]. In addition, an acute elevation in temperature, such as heat shock leads to fragmentation of both ER and Golgi [36]. In some mammalian cells and animal models mild elevation in temperature (up to 40°C) cause the development of thermotolerance, which is linked with increased expression of heat shock proteins and ER stressors [37, 38]. Moreover, mild hypothermia (28°C) stimulates mild ER stress in human pluripotent stem cells [39].

#### *4.2.5 Physiological ER stress signaling*

In physiological stress conditions, for instance increased secretory level or pathological stress, induced by aggregation of mutant protein could lead to imbalance between requirement for folded protein and ER potential to fold protein, consequently causing ER stress [40]. In response to stress, eukaryotic cells have developed signal transduction pathways, known as unfolded protein response (UPR) (**Figure 2**). These pathways are regulated by a group of proteins that sense ER stress. These proteins generate stress signals that protect cell from damage or induce cell apoptosis. A strong link between UPR signaling and human diseases has been reported [41].

The main objective of the UPR is to restore homeostasis and inhibit ER stress by employing the following mechanisms: (a) increasing protein folding ability through expression of protein-folding chaperones (b) Inhibition of protein translation by downregulating the ER protein load and facilitating the denaturation of misfolded proteins. But in case of acute stress the UPR stimulates programmed cell death [40]. UPR-mediated cell death is responsible for the onset of many diseases (**Table 1**), including cancer, type 2 diabetes, neurodegeneration, and atherosclerosis (40).

#### *4.2.6 Role of ER in metabolism*

The ER plays crucial role in the regulation of metabolic reactions. More specifically, the UPR pathway is involved in the regulation of glycolysis and it was recently reported that a regulatory protein mediates a metabolic decrease upon decrease in glucose level in neurons, suggesting an important role for the UPR as an adaptive response mechanism in relation to energy metabolism [42]. In addition, another signaling molecule known as mTOR maintains protein synthesis.

To maintain lipid content in the body ER plays an important role. Hepatocytes, liver cells carry SER in abundance, because along with protein synthesis, these cells also produce bile acids, cholesterol and phospholipids. Lipid accumulation leads to

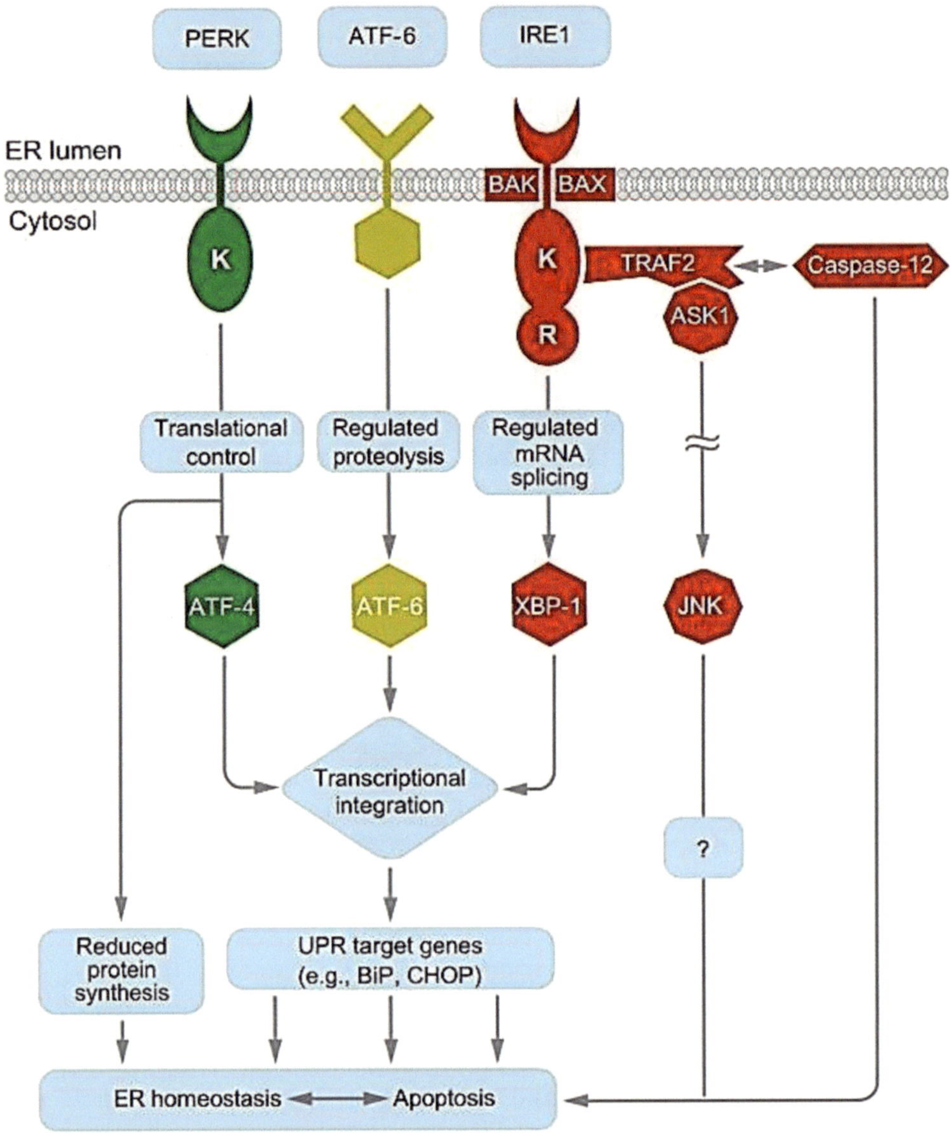


**Figure 2.**
*A schematic of mammalian unfolded protein response (UPR) signaling. IRE1, PERK, and ATF-6 proteins reside at the ER membrane. In response to ER stress, they initiate a cascade of signal transduction outputs that control cell survival or death.*

lipotoxicity, which is the fundamental cause of metabolic diseases. ER carry various enzymes that are crucial for lipid metabolism. Cellular cholesterol level is regulated via signaling pathways. One of these pathways is SCAP/SREBP2, which converts cholesterol into oxysterols and eventually to bile acids. Similarly, level of intracellular fatty acid is controlled by ER by a bunch of enzymes including, desaturases, elongases, and beta oxidation cycles [43].

The UPR has also been reported important for amino acid metabolism. Amino acid synthesis is closely linked with demand for protein biogenesis during ER stress. In response to ER stress, amino acid biosynthetic genes are expressed.

| Disease | Role of ER stress |
|---|---|
| Alzheimer's disease | Mutant presenilin 1 induces CHOP |
| Parkinson's disease | Accumulation of a substrate of Parkin in the ER activates ER stress |
| Amyotrophic lateral sclerosis | Mutant SOD1 aggregates and activates ER stress |
| Type 2 diabetes | Obesity induces ER stress<br>ATF6 interferes with gluconeogenesis<br>Free fatty acids and hyperglycemia induce beta cell death through CHOP |
| Atherosclerosis | Atherosclerosis-relevant stimuli induce macrophage death via CHOP<br>Oxidized phospholipids, hyperhomocysteinemia, and cholesterol loading induce endothelial and smooth muscle cell death via CHOP |
| Nonalcoholic fatty liver disease | ER stress induces SREBP-1c |
| HCV and HBV infection | HCV suppresses IRE1-XBP1 pathway |
| Alcoholic liver disease | Alcohol induces Grp78 and CHOP |

**Table 1.**
*Diseases related to endoplasmic reticulum (ER) stress.*

#### *4.2.7 Role of ER stress in age-related diseases*

According to current studies, the process of aging and age-related disorders are interlinked with ER stress response [44]. With aging, the normal cellular functions are declined, particularly slow degradation of chaperones, which results in increased aggregation of misfolded proteins [45]. These misfolded proteins accumulated in various organs of the body, such as in case of Alzheimer's disease (AD), an inflammatory neurodegenerative disease. In AD brains, ER stress responses have been observed, because ER is the site for synthesis of secretory and membranous proteins [46].

Aging disrupts the balance between UPR and pro-apoptotic signaling, leading to reduced protective response against ER-stress signaling [47]. Essential chaperons and enzymes, required for protein folding are functionally damaged with aging [48]. ER structure is also altered with aging. Hinds and McNelly reported dispersion in the highly organized ER cisternae [49].

Autophagy, a process activated by UPR system, remove the aggregation of misfolded proteins. Nevertheless, this process slows down with age, leading to neurodegeneration [50]. ER stress responses have been associated with certain neurodegenerative diseases, including Alzheimer's disease, Parkinson's disease, ALS and Huntington's disease. In neurodegenerative disorders, UPR activity is sustained, while apoptotic pathways are upregulated, encompassed by accumulation of aggregated proteins, hence neuronal death in old age [51].

It has been postulated that ER stress also plays role in the onset of metabolic disorders. For instance, in Type 2 Diabetes, a metabolic disease, caused by insulin resistance is regulated by various ER stress response mechanisms. The two main mechanisms that disrupt insulin activity are interlinked with ER stress [52].

## 5. Conclusion

ER is a complex and well-organized organelle, crucial for various cellular metabolic functions. ER homeostasis is maintained by a network of signaling pathways, collectively known as ER stress response, in order to deal with genetic, infectious and inflammatory stressors. With age, UPR, ER stress response mechanism, lost its activity thus less efficiently respond to these stressors. This results in onset of various age-related metabolic and neurodegenerative diseases. Advent of ER stress targeted therapeutics, particularly those improving protein folding and efficiency of associated regulatory mechanisms, promoting early detection of misfolded proteins could be useful in preventing and treating age-related disorders discussed in this chapter. Moreover, detection of anomalies in the ER stress response may led to development of therapeutics that could maintain ER homeostasis. This represents so far unexplored approach for disease prevention.

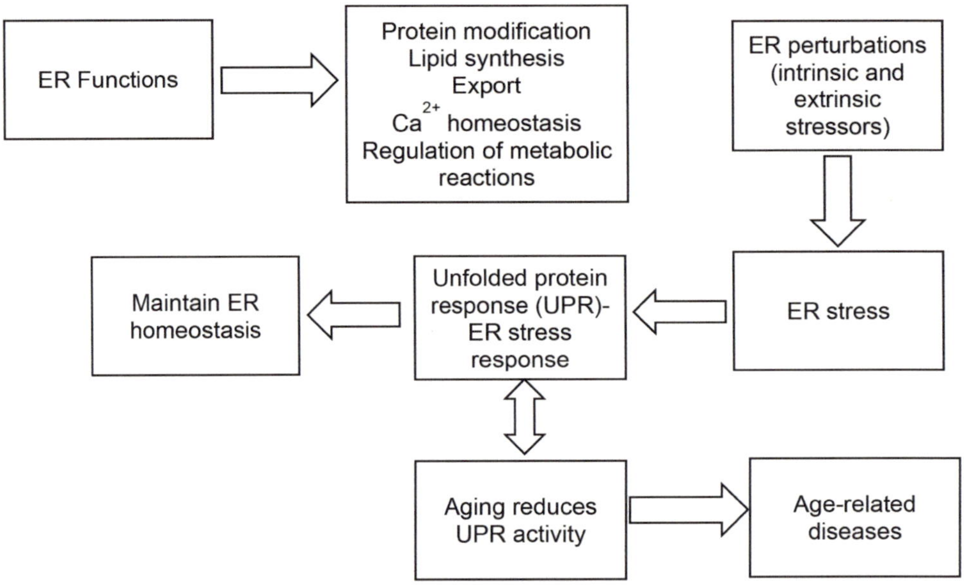


Schematic diagram summarizing chapter discussion

## Conflict of interest

The authors declare no conflict of interest.

## Author details

Sikander Ali[1*] and Maria Najeeb[2]

1 Institute of Industrial Biotechnology, GC University, Lahore, Pakistan

2 Institute of Microbiology, UVAS, Lahore, Pakistan

*Address all correspondence to: dr.sikanderali@gcu.edu.pk

## References

[1] Schwarz DS, Blower MD. The endoplasmic reticulum: Structure, function and response to cellular signaling. Cellular and Molecular Life Sciences. 2016;**73**(1):79-94

[2] Westrate LM, Lee JE, Prinz WA, Voeltz GK. Form follows function: The importance of endoplasmic reticulum shape. Annual Review of Biochemistry. 2015;**84**:791-811

[3] Alberts B, Johnson A, Lewis J, Raff M, Roberts K, Walter P. Molecular Biology of the Cell. 4th ed. New York: Garland Science; 2002

[4] Stefan CJ, Manford AG, Baird D, Yamada-Hanff J, Mao Y, Emr SD. Osh proteins regulate phosphoinositide metabolism at ER-plasma membrane contact sites. Cell. 2011;**144**:389-401

[5] Haustrate A, Prevarskaya N, Lehen'kyi VY. Role of the TRPV channels in the endoplasmic reticulum calcium homeostasis. Cells. 2020;**28**:317

[6] Hayashi T, Rizzuto R, Hajnoczky G, Su TP. MAM: More than just a housekeeper. Trends in Cell Biology. 2009;**19**:81-88

[7] Toulmay A, Prinz WA. Lipid transfer and signaling at organelle contact sites: The tip of the iceberg. Current Opinion in Cell Biology. 2011;**23**:458-463

[8] Wilhelm LP, Tomasetto C, Alpy F. Touche! STARD3 and STARD3NL tether the ER to endosomes. Biochemical Society Transactions. 2016;**44**:493-498

[9] Wilhelm LP, Wendling C, Vedie B, Kobayashi T, Chenard M, Tomasetto C, et al. STARD3 mediates endoplasmic reticulum-toendosome cholesterol transport at membrane contact sites. EMBO Journal. 2017;**36**:1412-1433

[10] Uemura T, Yamamoto M, Kametaka A, Sou Y, Yabashi A, Yamada A, et al. A cluster of thin tubular structures mediates transformation of the endoplasmic reticulum to autophagic isolation membrane. Molecular and Cellular Biology. 2014;**34**:1695-1706

[11] Braakman I, Bulleid NJ. Protein folding and modification in the mammalian endoplasmic reticulum. Annual Review of Biochemistry. 2011;**80**:71-99

[12] Meusser B, Hirsch C, Jarosch E, Sommer T. ERAD: The long road to destruction. Nature Cell Biology. 2005;7:766-772

[13] Hebert DN, Bernasconi R, Molinari M. ERAD substrates: Which way out? Semin Frontiers in Developmental and Cell Biology. 2010;**21**:526-532

[14] Lim CY, Davis OB, Shin HR, Zhang J, Berdan CA, Jiang X, et al. ER–lysosome contacts enable cholesterol sensing by mTORC1 and drive aberrant growth signalling in Niemann–Pick type C. Nature Cell Biology. 2019;**21**(10):1206-1218

[15] Jo Y, DeBose-Boyd RA. Control of cholesterol synthesis through regulated ER-associated degradation of HMG CoA reductase. Critical Reviews in Biochemistry and Molecular Biology. 2010;**45**:185-198

[16] Gorur A, Yuan L, Kenny SJ, Baba S, Xu K, Schekman R. COPII-coated membranes function as transport carriers of intracellular procollagen I. Journal of Cell Biology. 2017;**216**(6):1745-1759

[17] Siddiqi S, Saleem U, Abumrad NA, Davidson NO, Storch J, Siddiqi SA, et al. A novel multiprotein complex is required to generate the prechylomicron transport vesicle from intestinal ER. The Journal of Lipid Research. 2010;**51**:1918-1928

[18] Hetz C. The unfolded protein response: Controlling cell fate decisions under ER stress and beyond. Nature Reviews Molecular Cell Biology. 2012;**13**:89-102

[19] Hanahan D, Weinberg RA. Hallmarks of cancer: The next generation. Cell. 2011;**144**:646-674

[20] Holdereld M, Deuker MM, McCormick F, McMahon M. Targeting RAF kinases for cancer therapy: BRAF mutated melanoma and beyond. Nature Reviews Molecular Cell Biology. 2014;**14**:455-467

[21] Piwocka K, Vejda S, Cotter TG, O'Sullivan GC, McKenna SL. Bcr-Abl reduces endoplasmic reticulum releasable calcium levels by a Bcl-2independent mechanism and inhibits calcium-dependent apoptotic signaling. Blood. 2006;**107**:4003-4010

[22] Chen H-J, Anagnostou G, Chai A, Withers J, Morris A, Adhikaree J, et al. Characterization of the properties of a novel mutation in VAPB in familial amyotrophic lateral sclerosis. Journal of Biological Chemistry. 2010;**285**:40266-40281

[23] Nishimura AL, Mitne-Neto M, Silva HCA, RichieriCosta A, Middleton S, Cascio D, et al. A Mutation in the vesicle-trafficking protein VAPB causes late-onset spinal muscular atrophy and amyotrophic lateral sclerosis. The American Journal of Human Genetics. 2004;**75**:822-831

[24] Harding HP, Ron D. Endoplasmic reticulum stress and the development of diabetes: A review. Diabetes. 2002;**51**:S455-S461

[25] Oyadomari S, Koizumi A, Takeda K, Gotoh T, Akira S, Araki E, et al. Targeted disruption of the Chop gene delays endoplasmic reticulum stress-mediated diabetes. The Journal of Clinical Investigation. 2002;**109**:525-532

[26] Colombo C, Porzio O, Liu M, Massa O, Vasta M, Salardi S, et al. Seven mutations in the human insulin gene linked to permanent neonatal/infancy-onset diabetes mellitus. The Journal of Clinical Investigation. 2008;**118**:2148-2156

[27] Støy J, Edghill EL, Flanagan SE, Ye H, Paz VP, Pluzhnikov A, et al. Insulin gene mutations as a cause of permanent neonatal diabetes. The Proceedings of the National Academy of Sciences USA. 2007;**104**:15040-15044

[28] Giampietri C, Petrungaro S, Conti S, Facchiano A, Filippini A, Ziparo E. Cancer microenvironment and endoplasmic reticulum stress response. Mediators of Inflammation. 2015;**2015**:417281

[29] Jin ML, Park SY, Kim YH, Oh J-I, Lee SJ, Park G. The neuroprotective effects of cordycepin inhibit glutamate-induced oxidative and ER stressassociated apoptosis in hippocampal HT22 cells. Neurotoxicology. 2014;**41**:102-111

[30] Kim SH, Kwon DY, Kwak JH, Lee S, Lee YH, Yun J, et al. Tunicamycin-induced ER stress is accompanied with oxidative stress via abrogation of sulfur amino acids metabolism in the liver. International Journal of Molecular Sciences. 2018;**19**(12):4114

[31] Xiang XY, Yang XC, Su J, Kang JS, Wu Y, Xue YN, et al. Inhibition of autophagic flux by ROS promotes apoptosis during DTT-induced ER/

oxidative stress in HeLa cells. Oncology Reports. 2016;**35**(6):3471

[32] Gallagher CM, Walter P. Ceapins inhibit ATF6α signaling by selectively preventing transport of ATF6α to the Golgi apparatus during ER stress. Elife. 2016;**20**(5):e11880

[33] Malo A, Krcuger B, Gcoke B, Kubisch CH. 4phenylbutyric acid reduces endoplasmic reticulum stress, trypsin activation, and acinar cell apoptosis while increasing secretion in rat pancreatic acini. Pancreas. 2013;**42**:92-100

[34] Lee YY, Hong SH, Lee YJ, Chung SS, Jung HS, Park SG, et al. Tauroursodeoxycholate (TUDCA), chemical chaperone, enhances function of islets by reducing ER stress. Biochemical and Biophysical Research Communications. 2010;**397**:735-739

[35] Ozcan U, Yilmaz E, Ozcan L, Furuhashi M, Vaillancourt E, Krcuger B, et al. Chemical chaperones reduce ER stress and restore glucose homeostasis in a mouse model of type 2 diabetes. Science. 2006;**313**:1137-1140

[36] Lepock JR. How do cells respond to their thermal environment? International Journal of Hyperthermia. 2005;**21**:681-687

[37] Bettaieb A, Averill-Bates DA. Thermotolerance induced at a mild temperature of 40°C protects cells against heat shock-induced apoptosis. Journal of Cellular Physiology. 2005;**205**:47-57

[38] Liu Y, Sakamoto H, Adachi M, Zhao S, Ukai W, Hashimoto E, et al. Heat stress activates ER stress signals which suppress the heat shock response, an effect occurring preferentially in the cortex in rats. Molecular Biology Reports. 2005;**39**:3987-3993

[39] Mollereau B. Cooling-induced ER stress is good for your brain. EBioMedicine. 2015;**2**:482-483

[40] Tabas I, Ron D. Integrating the mechanisms of apoptosis induced by endoplasmic reticulum stress. Nature Cell Biology. 2011;**13**:184-190

[41] Marciniak SJ, Ron D. Endoplasmic reticulum stress signaling in disease. Physiological Reviews. 2006;**86**:1133-1149

[42] Van der Harg JM, van Heest JC, Bangel FN, Patiwael S, van Weering JRT, Scheper W. The UPR reduces glucose metabolism via IRE1 signaling. Biochimica et Biophysica Acta. 2017;**1864**:655-665

[43] Zhou H, Liu R. ER stress and hepatic lipid metabolism. Frontiers in Genetics. 2014;**5**:112

[44] Höhn A, Tramutola A, Cascella R. Proteostasis failure in neurodegenerative diseases: Focus on oxidative stress. Oxidative Medicine and Cellular Longevity. 2020;**27**:2020

[45] Macario AJL, De Macario EC. Sick chaperones and ageing: A perspective. Ageing Research. 2002;**1**:295-311

[46] Uddin M, Tewari D, Sharma G, Kabir M, Barreto GE, Bin-Jumah MN, et al. Molecular mechanisms of ER stress and UPR in the pathogenesis of Alzheimer's disease. Molecular Neurobiology. 2020;**57**(7):2902-2919

[47] Naidoo N, Ferber M, Master M, Zhu Y, Pack AI. Aging impairs the unfolded protein response to sleep deprivation and leads to proapoptotic signaling. Journal of Neuroscience. 2008;**28**:6539-6548

[48] Nuss JE, Choksi KB, DeFord JH, Papaconstantinou J. Decreased enzyme

activities of chaperones PDI and BiP in aged mouse livers. Biochemical Biophysical Research Communications. 2008;**365**:355-361

[49] Hinds J, McNelly N. Dispersion of cisternae of rough endoplasmic reticulum in aging CNS neurons: A strictly linear trend. The American Journal of Anatomy. 2005;**152**:433-439

[50] Doyle KM, Kennedy D, Gorman AM, Gupta S, Healy SJ, Samali A. Unfolded proteins and endoplasmic reticulum stress in neurodegenerative disorders. Journal of Cellular and Molecular Medicine. 2011;**15**:2025-2039

[51] Xiang C, Wang Y, Zhang H, Han F. The role of endoplasmic reticulum stress in neurodegenerative disease. Apoptosis. 2017;**22**(1):1-26

[52] Hotamisligil GS. Endoplasmic reticulum stress and atherosclerosis. Nature Medicine, Nature Medicine. 2010;**16**:396-399

Chapter 2

# Endoplasmic Reticulum: A Hub in Lipid Homeostasis

*Raúl Ventura and María Isabel Hernández-Alvarez*

## Abstract

Endoplasmic Reticulum (ER) is the largest and one of the most complex cellular structures, indicating its widespread importance and variety of functions, including synthesis of membrane and secreted proteins, protein folding, calcium storage, and membrane lipid biogenesis. Moreover, the ER is implicated in cholesterol, plasmalogen, phospholipid, and sphingomyelin biosynthesis. Furthermore, the ER is in contact with most cellular organelles, such as mitochondria, peroxisomes, Golgi apparatus, lipid droplets, plasma membrane, etc. Peroxisomes are synthesized from a specific ER section, and they are related to very-long-chain fatty acid metabolism. Similarly, lipid droplets are vital structures in lipid homeostasis that are formed from the ER membrane. Additionally, there is a specific region between the ER-mitochondria interface called Mitochondria-Associated Membranes (MAMs). This small cytosolic gap plays a key role in several crucial mechanisms from autophagosome synthesis to phospholipid transfer. Due to the importance of the ER in a variety of biological processes, alterations in its functionality have relevant implications for multiple diseases. Nowadays, a plethora of pathologies like non-alcoholic steatohepatitis (NASH), cancer, and neurological alterations have been associated with ER malfunctions.

**Keywords:** endoplasmic reticulum, mitochondria, peroxisomes, mitochondria, lipid droplets, MAM, phospholipid, lipid metabolism

## 1. Introduction

The endoplasmic reticulum (ER) is a dynamic organelle largely responsible for essential cellular functions. Its wide and diverse functionality transforms the ER into a key organelle in cellular stress, signaling, vesicle transport, and lipid homeostasis. The ER is often in a state of constant change, shifting its structure to promote cell adaptation to environmental changes. For this reason, ER mass or area can fluctuate depending on cellular state and conditions.

The ER membrane is a lipid bilayer comprising two compartments: a cytosolic region in contact with the cytoplasm and a luminal region, which is the space between the two ER membranes.

In addition, two very well differentiated structures can be found within the ER: the rough endoplasmic reticulum (RER) and the smooth endoplasmic reticulum (SER). These structures have a unique architecture that is specialized for different cellular mechanisms. Specifically, RER is formed of flattened sheets and contains a large quantity of ribosomes, whereas SER has a more irregular construct, consisting of a tubular structure, and is lacking in ribosomes [1–3]. This indicates that RER is more associated with protein synthesis than SER and newly synthesized proteins in RER ribosomes could enter the RER lumen to achieve their final conformation. Therefore, in cell types that hold a high rate of protein synthesis, such as hepatocytes, it is essential that the RER is further developed [1, 4].

In contrast, SER has additional functions related to calcium storage, detoxification, lipid metabolism, steroidal hormones, and bile acid production. As a result, SER is more developed in cells with a high detoxification power, such as hepatocytes, or in both smooth and skeletal muscle cells (where SER is called sarcoplasmic reticulum). Further, since SER participates in lipid metabolism, its presence is required in adipocytes as well [1, 2, 5].

Additionally, there is a specific region within SER and RER called nanodomains, where molecules and proteins are grouped, harmonically working together to perform a precise function. Thus, proteins are not evenly distributed along with the ER but associated in clusters [6].

As mentioned, RER is specialized to accommodate both membrane and secreted protein synthesis, as well as post-translational modifications. In particular, this ER region is in contact with the nuclear envelope and allows the mRNA to enter from the nucleus to the RER lumen. The RER's involvement in post-translational modifications includes glycosylation (N-glycosylation), sulfurization, and correct protein folding. First, RER-directed proteins must be recruited via a specific signal peptide which is recognized by the ribonucleoprotein complex SRP (Signal Recognition Particle). Once this target signal is recognized, the protein synthesis is temporarily stopped, and the ribosome, mRNA, and the newly synthesized protein are transported to the RER membrane. Here, the complex interacts with a membrane receptor and the protein is translocated to the RER lumen. Nonetheless, the mRNA continues to be translated and, at this point, protein synthesis recommences. Finally, the process ends when the entirety of the protein is in the RER lumen, and its signal peptide is cleaved by a peptidase. In the lumen, some chaperones are present in order to avoid an incorrect folding of the newly synthesized protein. There are also a plethora of enzymes that modify proteins in this region, preparing them to be secreted, for instance [4, 7–10].

The ER has a "quality-control activity" that detects misfolded proteins. If the proteins cannot be successfully repaired, they are discarded and degraded. Specifically, misfolded proteins are detected by glucosyltransferase, which binds glucose to them. The bound glucose is recognized by calnexin, which attempts to correctly fold the protein on several occasions. When the problem is not solved, the misfolded protein is degraded through the proteasome, yielding amino acids that the cell can recycle into new proteins. However, when misfolded proteins reach relevant levels, they are detected by some sensors within the ER membrane. These sensors regulate cellular responses to ER stress in a process called UPR or unfolded protein response, and can imply both survival and non-survival pathways [11–13].

Despite the variety of functions of the ER, this chapter will primarily focus on two aspects: the ER's implication in lipid metabolism, transport, synthesis, and homeostasis, and the mechanism by which the endoplasmic reticulum interacts with other organelles to achieve this.

As stated above, the ER has specific regions of contact, termed membrane contact sites (MCS), with a multitude of cellular organelles. However, it should be noted that these interactions do not imply membrane fusion but classically they have been shown to be related to calcium flux regulation (**Figure 1**).

*DOI: http://dx.doi.org/10.5772/intechopen.105450*

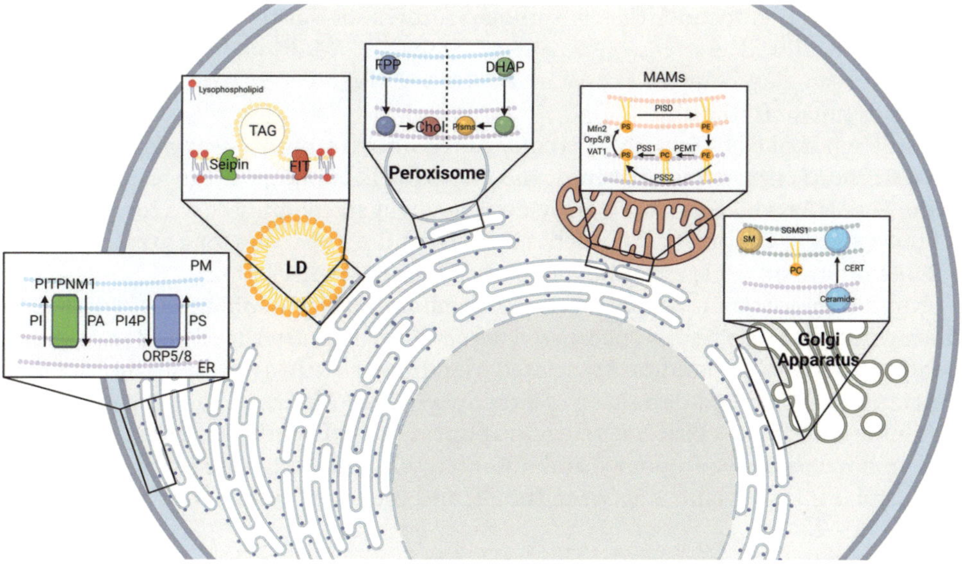


**Figure 1.**
*Graphical scheme of endoplasmic reticulum interactions with peroxisomes, mitochondria, lipid droplets, plasma membrane, and Golgi apparatus. With mitochondria, ER works coordinately to synthesize some of the major phospholipids, including phosphatidylcholine (PC), phosphatidylserine (PS), and phosphatidylethanolamine (PE). Contacts between ER and peroxisomes permit the synthesis of plasmalogen and cholesterol, while sphingomyelin (SM) is obtained as a result of interactions between ER and Golgi apparatus. Both peroxisomes and lipid droplets derive from ER membranes, implicating proteins such as Seipin and FIT (in lipid droplet biogenesis), or Pex3 and Pex19 (in peroxisomes biogenesis). However, ER can also interact with more organelles such as lysosomes. ER can also contact with plasma membrane; these contacts permit, for instance, the phosphatidylinositol (PI), phosphatidic acid (PA), PS, and phosphatidylinositol 4-phosphate (PI4P) transport, thanks to PITPNM1 and ORP5/8 action.*

Nevertheless, at present, MCS is also seen to be associated with lipid exchange, lipid synthesis (phospholipids, sphingomyelin, cholesterol, or plasmalogen biosynthesis, for example), and vesicle traffic [3, 14–16].

## 2. Endoplasmic reticulum and mitochondria

Mitochondria are known to be the powerhouse of the cell due to their energy production and homeostasis-related functions. For instance, some of the main processes in ATP obtention, such as the respiratory chain reactions or the Krebs cycle, occur in the mitochondria. Hence, mitochondria are especially essential in organs, tissues, or cell structures that require profuse amounts of energy, including the heart, neurons, or sperm flagella.

Interestingly, mitochondria can also enhance programmed cell death, apoptosis. Particularly, it intervenes in the intrinsic or cellular pathway, which can be activated by different cellular stimuli, including DNA damage, or ER stress. The process begins with the liberation of pro-apoptotic substances such as Cytochrome C, which participates in the respiratory chain. Cytochrome C then goes on to activate a signal cascade leading to cellular death. Clearly, mitochondria have an important involvement in some vital cellular functions, meaning its dysfunction is implicated in a large number of diseases, such as cancer, metabolic, neuronal, cardiovascular, or genetic disorders [17, 18].

Furthermore, mitochondria have unique features such as the presence of their own circular DNA as well as an inner and outer membrane. Additionally, they are highly dynamic organelles, being in a constant fusion and fission cycle based on cellular state and environmental stimuli.

Mitochondrial DNA (mtDNA) is constituted of two strands, heavy and light. The heavy strand is enriched in guanines, and codes for 12 subunits of the respiratory chain, 14 tRNAs, and two rRNAs. However, the light strand only codes for 8 tRNAs and one subunit. Despite that mitochondrial DNA does not code for a great quantity of RNAs or proteins, it is essential for good cell functionality [19–21].

Another important role attributed to mitochondria is phospholipid biogenesis, which takes place in a highly specialized association between the ER and the mitochondria called Mitochondria-Associated Membranes (MAM). Despite the fact there is currently no specified definition of exactly what MAMs are, it has been established that this area regulates processes such as apoptosis, lipid synthesis, transport, calcium homeostasis, autophagy, and mitophagy. MAMs structure is also required for phospholipids transport between the ER and the outer mitochondrial membrane (OMM) [15, 22–24].

Calcium is an essential regulatory element in mitochondria that regulates metabolism, apoptosis, and autophagy. In MAMs, calcium ion transfer to organelles is promoted due to its abundance of calcium transport channels. Specifically, mitochondria uptakes calcium ions through outer membrane voltage-dependent anion channels (VDAC) [25].

### 2.1 Phospholipid synthesis: ER and mitochondria cooperation

As previously explained, the endoplasmic reticulum and mitochondria coordinate together to synthesize some of the most important glycerophospholipids, such as phosphatidylcholine (PC), phosphatidylethanolamine (PE), and phosphatidylserine (PS), which represents approximately 50% of total phospholipids located in cellular membranes [26].

Glycerophospholipids are the major component in cellular membranes, both intracellular (ER, mitochondrial, and peroxisomal membranes) and extracellular facing (plasma membrane). These lipids present a polar head (phosphate group) and a hydrophobic tail, formed by fatty acids. This duality gives them an amphipathic character, endowing their characteristics to the membranes. Moreover, the hydrophobic tail can be bound with choline, ethanolamine, and serine forming, PC, PE, and PS, respectively [27, 28].

In mammals, all glycerophospholipids are synthesized from a common molecule, diacylglycerol (DAG), which derives from phosphatidic acid. Throughout the synthesis process, a large number of enzymes and many intermediate molecules are generated. The vast majority of these molecules are in the ER or the mitochondrial membrane. The first step in glycerophospholipid synthesis is phosphatidic acid generation. Two acyltransferases (glycerol-3-phosphate acyltransferase-1, GPAT1, and acylglycerophosphate acyltransferase) located in the ER and outer mitochondrial membrane must act on a glycerol-3-P molecule. The phosphatidic acid-phosphatase 1 (PAP-1), a cytosolic enzyme activated upon contact with ER membrane, acts on phosphatidic acid-generating DAG. Finally, PC, PS, and PE are synthesized from DAG [29, 30].

Mammalian cells can synthesize phosphatidylethanolamine from phosphatidylserine. In order for this to occur, PS needs to be decarboxylated by the action of an inner mitochondrial membrane enzyme, PISD (mitochondrial phosphatidylserine decarboxylase). For PE formation, an alternative synthesis pathway is utilized, the Kennedy pathway. In this pathway, ethanolamine kinase phosphorylates ethanolamine, that comes from the extracellular environment via plasma membrane, and there is several intermediate enzymatic steps, leading to the formation of PE in the ER [31].

The Kennedy pathway is also used to obtain phosphatidylcholine. When choline is in the cytoplasm, it is phosphorylated by the choline kinase. Once phosphorylated, the choline-phosphate cyticylyltransferase catalyzes CDP-choline (cytidine-5-diphosphocholine) formation. Afterward, 1,2-diacylglycerol choline phosphotransferase, transfers a DAG molecule to CDP-choline, finally generating phosphatidylcholine in the ER [32].

In parallel, PE can also be converted to PC, but PE must first be methylated three times by phosphatidylethanolamine N-methyltransferase (PEMT), which is located in ER membrane. In general, this is not a representative pathway, except for in hepatocytes, where there are significant quantities of phosphatidylcholine produced [32, 33].

There are two enzymes found in MAMs that can synthesize phosphatidylserine: PSS1 (Phosphatidylserine Synthase 1) and PSS2 (Phosphatidylserine Synthase 2). PSS1 catalyzes the exchange of choline from phosphatidylcholine for serine, whereas PSS2 performs the equivalent exchange with ethanolamine from phosphatidylethanolamine. In both cases, phosphatidylserine is obtained [34, 35].

Recently, it has been described that Mitofusin 2 (Mfn2) participates in the phosphatidylserine transport between ER and mitochondrial outer membrane. Mfn2 is a GTPase protein located in the outer mitochondrial membrane, that classically, was associated with the process of mitochondrial fusion, regulating the fusion of two OMM [36].

Beyond controlling the mitochondrial fusion process, *Hernández-Alvarez et al.* demonstrated that the ablation of Mfn2 in mouse livers causes inflammation, triglyceride accumulation, fibrosis, and liver cancer with age. In addition, a reduction of Mfn2 levels has been observed in hepatic biopsies from patients with non-alcoholic

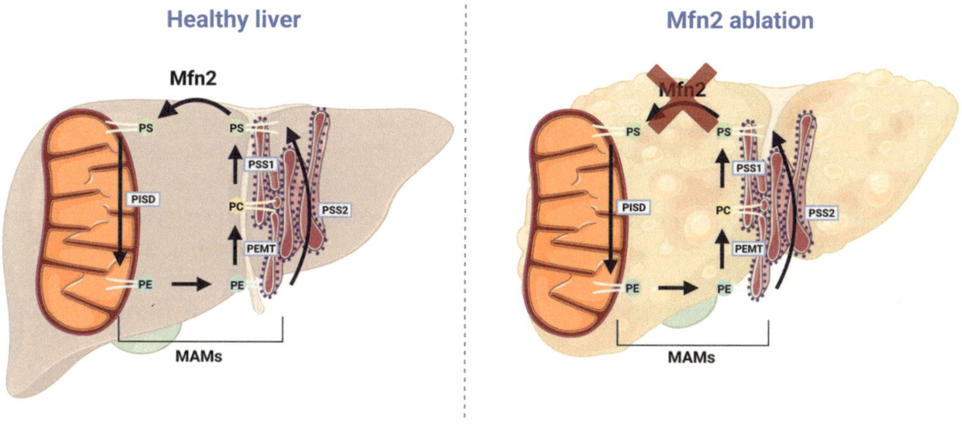


**Figure 2.**
*Graphical scheme of phosphatidylcholine (PC), phosphatidylserine (PS), and phosphatidylethanolamine (PE) biosynthesis in mitochondria-associated membranes (MAMs) in a healthy (left) and Mitofusin 2-ablated (right) liver. In ER, PC thanks to the phosphatidylserine synthase 1 (PSS1), is transformed to PS, which is transported to mitochondria due to Mitofusin 2 (Mfn2) action. There, PS is converted to PE, a process catalyzed by phosphatidylserine decarboxylase (PISD), which is in the mitochondrial membrane. Once PE is synthesized, it is transferred to ER, where is converted to PC or PS, depending on the enzyme implicated, phosphatidylethanolamine N-methyltransferase (PEMT) or phosphatidylserine synthase 2 (PSS2), respectively.*

steatohepatitis (NASH), a disease related to lipid metabolism. The levels of this protein were also lower in mouse models of steatosis or NASH. Furthermore, its re-expression in a NASH mouse model ameliorated the disease, therefore, demonstrating that Mfn2 protects against liver disease [15].

A probable explanation for the protective role of Mfn2 is presented in the same study; the authors demonstrated that Mfn2 has the ability to bind and help in the transfer of phosphatidylserine across ER-mitochondria contacts, generating PS-enriched domains. This facilitates PS transfer to mitochondria and further mitochondrial phosphatidylethanolamine synthesis. This transfer occurs in MAMs (**Figure 2**). Hence, a reduction of Mfn2 hepatic levels leads to poor PS transfer and phospholipid synthesis, causing ER stress, NASH-like phenotype, and liver cancer [15].

Additionally, Mfn2 deficiency alters PSS1 and PSS2 protein levels, inhibiting PS synthesis. This was also observed in the Mfn2 liver knock-out mouse model. The lack (or reduction) of Mfn2 also generates MAMs remodeling (altering the phospholipid composition in ER-mitochondrial contact sites). This modification leads to triglyceride accumulation, insulin resistance, and impaired phospholipid synthesis [15]. Other proteins associated with PS transport are oxysterol binding related proteins 5 and 8 (ORP 5/8) and synaptic vesicle membrane protein (VAT1) [24, 37]. ORP8 downregulation was also related to liver cancer [38]. However, whether PS deficiency is the cause, or the consequence needs to be further investigated.

## 3. Endoplasmic reticulum and peroxisomes

Peroxisomes are a highly versatile single membrane organelle present in many eukaryotic cells, including yeast. They can modify their morphology, size (0.2–1.5 μm), number, and activity depending on their nutritional state, cell type, or cellular environment. In mammals, peroxisomes contain a diverse range of enzymes making them organelles essential for several biochemical pathways. Some of the many roles of the peroxisome include fatty acid β-oxidation, bile acid synthesis, amino acid catabolism, polyamine oxidation, metabolism of reactive oxygen, and nitrogen species. Though all these functions are relevant, fatty acid β-oxidation is the most relevant. This process is critical for very-long-chain fatty acids (VLCFA) shortening, which mitochondria are not able to metabolize [39–41].

There are large amounts of oxidative enzymes contained within peroxisomes which can be observed in electron microscopy as crystals inside the organelle. Some of these enzymes include oxidase and catalase which use molecular oxygen to oxidize fatty and amino acids. Due to the high grade of toxicity within the peroxisome, catalase uses hydrogen peroxide ($H_2O_2$) to eliminate toxic/harmful substances, such as ethanol or methanol, or to oxidize new substrates. For this reason, peroxisomes are more abundant in cells undergoing detoxification processes, such as hepatocytes or kidney cells [39–42].

It has been described that peroxisomes interact with different organelles (lipid droplets, ER, mitochondria, lysosomes, etc.) through their contact sites to maintain lipid homeostasis and metabolism. For instance, peroxisomes transform VLCFA into medium-chain fatty acids, lipids that will be converted into water and $CO_2$ by mitochondrial action. Alterations in peroxisomes are associated with several pathologies and rare genetic diseases, usually affecting the brain, kidney, liver, and skeletal muscle. In the brain, peroxisomes play a crucial role in the synthesis of plasmalogens, a phospholipid especially enriched in myelin. Any alteration in peroxisomes will lead to severe demyelination in neurons, causing the neurological component observed in peroxisomal diseases. One example is the Zellweger syndrome, produced by a deficiency in peroxisomal biogenesis [39, 40, 42–44].

### 3.1 Peroxisome biogenesis: ER role

The peroxisomal membrane has a similar composition to the endoplasmic reticulum membrane. This provides a good understanding of the origin of peroxisomes; pre-peroxisomal vesicles in the ER (a specialized and delimited area of the ER). However, peroxisomes can also derive from another peroxisome through a process of fission [45]. During peroxisomal budding, a number of peroxisomal membrane proteins (PMP) are first directed to the ER, specifically in the specialized ER subdomain, from which pre-peroxisomes are budded. Pex3 and Pex19 are proteins that are particularly relevant in this process; Pex3 is a PMP located in ER membrane, while Pex19 is found in the cytosol. Pre-peroxisomal budding occurs due to an interaction between these two proteins [45, 46]. As the ER participates in peroxisomal synthesis and peroxisomes play a relevant role in lipid metabolism (plasmalogen and cholesterol synthesis), it could be suggested that the ER has an indirect function in all these processes.

### 3.2 Peroxisome-ER coordination in lipid homeostasis: cholesterol and plasmalogens biosynthesis

ER and peroxisomes work in coordination to maintain lipid homeostasis. There is an intimate relationship between these two organelles, transferring components and essential molecules to each other. For example, the ER transfers some important lipids to peroxisomes, whilst the ER receives some plasmalogens precursors from the peroxisome. Plasmalogens are ether phospholipids that represent approximately 20% of total phospholipids in humans. Synthesis of these molecules begins in peroxisomes and ends in the ER. As well as plasmalogens, peroxisomes can also synthesize cholesterol and part of its precursors. These precursors will then be transferred to the ER to complete their synthesis, demonstrating the complementary relationship between the ER and peroxisomes in their ability to synthesize cholesterol [39, 47, 48].

A major site of plasmalogens is in the nervous and immune system and heart; their main function is to protect these systems from oxidative damage produced by reactive oxygen species (ROS) or reactive nitrogen species (RNS). Synthesis of these molecules begins with the peroxisome phase, which initially involves esterification of dihydroxyacetone phosphate (DHAP) with an Acyl-CoA, catalyzed by DHAP acyltransferase (DHAP-AT). The resulting molecule, 1-acyl-DHAP, is then transformed to 1-O-alkyl-DHAP as a result of the action of alkyl dihydroxyacetone phosphate synthase (ADHAP-S), incorporating fatty alcohol and generating a fatty acid. Once 1-O-alkyl-DHAP is synthesized, it is transported to ER, where it will be transformed several times to obtain choline or ethanolamine plasmalogens (**Figure 3**). Hence, plasmalogen synthesis is another clear example of the relationship between peroxisomes and ER [39, 49, 50].

Nevertheless, the synthesis of this type of phospholipid is not the only process in which peroxisomes and the ER work together; a similar mechanism occurs with cholesterol. Principally, peroxisomes partake in the synthesis of farnesyl-pyrophosphate (FPP), an intermediate of terpenoid, terpene, and sterol biosynthesis; subsequently, the FPP generated is again transferred to the ER. Here, FPP experiences sequential modifications and finally results in the formation of cholesterol. Lastly, another means by which cholesterol synthesis can be initiated via peroxisomes is through

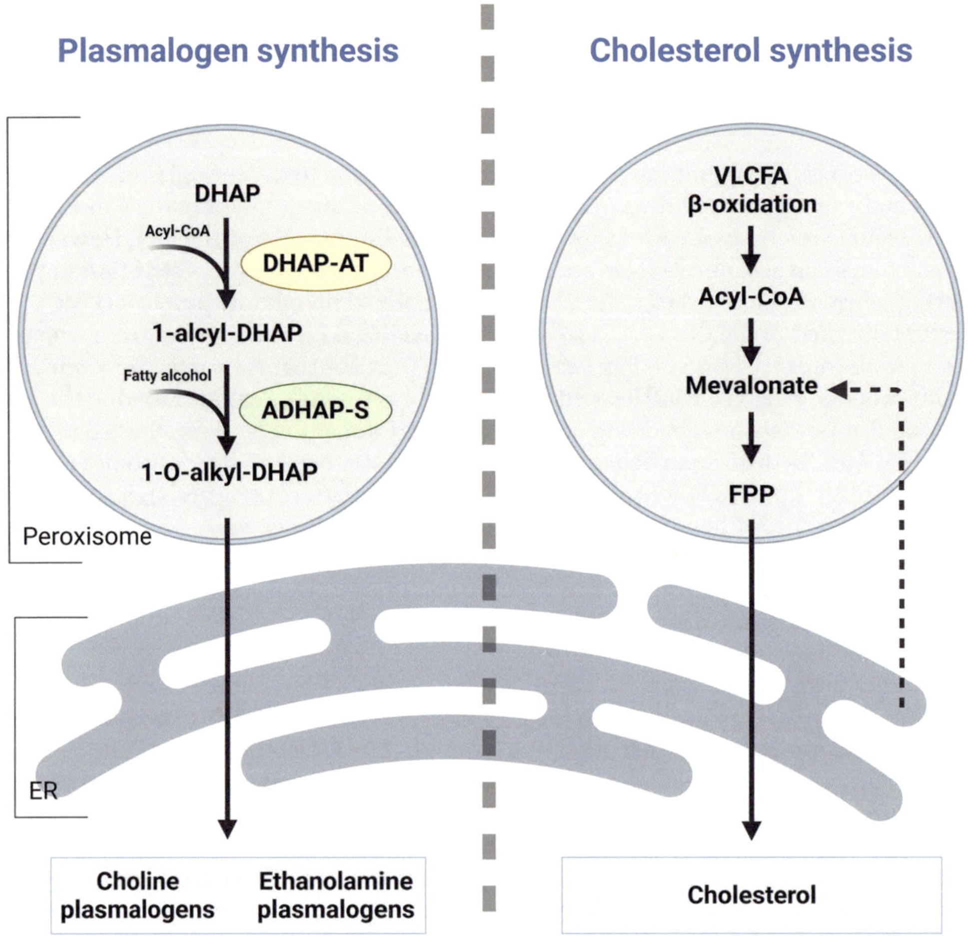


**Figure 3.**
*Graphical scheme of plasmalogen (left) and cholesterol (right) biosynthesis in peroxisome and endoplasmic reticulum. In plasmalogen synthesis, first, dihydroxyacetone phosphate (DHAP) is converted to 1-acyl-DHAP, a reaction catalyzed by DHAP acyltransferase (DHAP-AT), which incorporates an acyl-CoA molecule. Then, 1-O-alkyl-DHAP is obtained from 1-acyl-DHAP, through the incorporation of a fatty alcohol thanks to alkyl dihydroxyacetone phosphate synthase (ADAP-S). 1-O-alkyl-DHAP is transported to ER; there it suffers some modifications (not shown) until obtaining choline or ethanolamine plasmalogens. In cholesterol biosynthesis, the peroxisomal phase generates farnesyl-pyrophosphate (FPP), which is then transported to ER. There, FPP is modified (not shown), obtaining cholesterol. Mevalonate needed for FPP synthesis can be transferred from ER or be generated from acyl-CoA obtained during β-oxidation of very-long-chain fatty acids (VLCFA).*

Acetyl-CoA, derived from peroxisomal β-oxidation of very-long-chain fatty acids. Alternatively, it can continue synthesis from an intermediate, mevalonate, transferred from ER (**Figure 3**). In both cases, once FPP is obtained, cholesterol synthesis continues in the ER [47, 51, 52].

## 4. Endoplasmic reticulum and Golgi apparatus

Golgi apparatus is an organelle with two main functions: post-translational protein modification and sorting, packing, routing, and recycling membrane proteins. In the Golgi complex, four regions can be distinguished: ER-Golgi intermediate complex, cis and trans-Golgi network (the nearest and farthest cisternae to ER, respectively), and Golgi stack, which is divided into medial and trans compartments (corresponding to the central region of Golgi apparatus). Additionally, the Golgi complex has unique, biochemically distinct enzymes, that are distributed throughout its space [53, 54].

The newly synthesized proteins enter into the ER, where they are introduced into vesicles (they move from the ER-Golgi intermediate compartment to the cis-Golgi network). Finally, the vesicles reach the trans-Golgi network, which delivers these molecules to their target destinations. Despite the Golgi complex playing a fundamental role in protein transport and post-translational modifications, it is also involved in the synthesis of certain lipids, such as sphingolipids, and especially sphingomyelin, which is vital for correct cell functionality [53–55].

### 4.1 Golgi apparatus-ER phospholipid transport: sphingomyelin synthesis

Sphingomyelin is mainly located in the outer monolayer of the plasma membrane and is crucial for the functioning of a number of cellular processes, such as immune recognition, cell differentiation, growth, and apoptosis. Furthermore, sphingomyelin is known to be a major component of the myelin covering certain neuron axons. Namely, it binds hydrocarbon chains, improving myelin strength [56].

It is synthesized largely from ceramide and phosphatidylcholine, which are lipids obtained in the ER. Sphingomyelin synthase-1 (SGMS1), the enzyme required to synthesize sphingomyelin, is localized in the Golgi apparatus, thus its precursors must be transported from the ER to the Golgi complex.

Sphingomyelin can be obtained both in the plasma membrane and in the Golgi apparatus. Nonetheless, sphingomyelin synthesis is residual in the plasma membrane and this reaction is catalyzed by a different enzyme, sphingomyelin synthase-2 (SGMS2) [30, 57, 58].

As mentioned before, ceramide and phosphatidylcholine must be transferred from ER to the Trans Golgi network (TGN) to synthesize sphingomyelin, a process that occurs in the MCS. Ceramides are primarily located in ER membrane due to their low hydrophilicity. Moreover, this lipid can be transported to Golgi apparatus through one of two mechanisms: coatomer-dependent vesicular transport or action of a cytosolic peptide, the ceramide transfer protein (CERT). CERT transport is regulated by phosphatidylinositol-4-phosphate (PtdIns(4)P) quantity in the TGN. Once sphingomyelin is synthesized, it is transported to the plasma membrane via vesicular transport [30, 57, 58].

## 5. Endoplasmic reticulum and lipid droplets

Lipid droplets (LD), also known as adiposomes, are a spherical cytosolic organelle that stores triglycerides and cholesterol esters, providing an energy reserve. They are found in animals, plants, fungi, and in some bacteria. They comprise two different structures: a hydrophobic core formed by neutral lipids, and a polar phospholipid monolayer containing proteins (such as perilipin, PLIN), which partially regulate their functions. This ER-derived organelle plays a considerable role in lipid and energy homeostasis. For example, lipid droplets can generate contacts with a multitude of organelles, including mitochondria, peroxisomes, lysosomes, and the ER, allowing the transfer of lipids between them [59–61].

LD also seem to have a relevant role in infections where in some viral, fungal, or bacterial infections, microorganisms use LD in their cycle of infection. Examples that illustrate this can be seen in the case of some viruses, which exploit lipid droplets to assemble inside the cell or, on the other hand, mycobacteria and other intracellular pathogens, which steal the lipids contained in these structures to adapt themselves to the cellular environment [62]. As well as this, LD can also regulate cellular toxicity, accumulate toxic lipids, and protect vital structures from oxidative stress. Therefore, alterations in their function or physiology trigger diseases, such as NASH, obesity, or diabetes [59, 63, 64].

Depending on the cellular type, lipid droplets present distinct functionality. For instance, in adipocytes these droplets store triglycerides, waiting to be hydrolyzed when peripheral tissues require more energy. In testis, ovaries, or suprarenal glands, the lipid droplets are smaller than adipose tissue and they accumulate cholesterol esters, necessary for steroid and sex hormone synthesis. Finally, in the liver, adiposomes facilitate lipoprotein synthesis (low-density lipoprotein, LDL, and high-density lipoprotein, HDL). In mammals, tissues such as the liver, adipose tissue, and muscle contain an abundance of lipid droplets.

In addition, lipid droplet quantity is tightly associated with the nutritional status, metabolism, and nutrient availability of individuals. When there is an excess of nutrients, they are stored in lipids, increasing the number and size of lipid droplets. However, during starvation or nutrient depletion, lipids are mobilized, to synthesize the essential phospholipids or produce the required energy. Thus, lipid droplets are smaller and less abundant during this situation, especially in white adipose tissue and the liver [59, 60, 64, 65].

### 5.1 Lipid droplets biogenesis: ER role

As already explained, lipid droplets derive from a specific area of the ER membrane. Not only does the ER regulate and allow their genesis, but also has an important implication on lipid storage and metabolism (**Figure 4**).

Lipid droplets mainly store triglycerides and sterol esters, which are synthesized by enzymes mostly located in ER membrane. Prior to its incorporation into the lipid droplet, fatty acids are esterified with a sterol (to obtain a sterol ester) or diacylglycerol (to obtain a triglyceride). These reactions are catalyzed by acyl-CoA: cholesterol O-acyltransferases (ACAT1 and ACAT2), and diacylglycerol acyltransferases (DGAT1 and DGAT2), respectively. When the quantity of these lipids is significantly high, they tend to be grouped, forming what is called a "lipid lens" in the ER bilayer. It is thought that there are no proteins related to the formation of this lipid droplet precursor, due to a purely physical effect driven by their hydrophobicity characteristic. As more neutral lipids are synthesized, the lipid lenses tend to expand, eventually causing the lipid droplet to bud from ER membrane [66–69].

Unlike in secretory vesicles, lipid droplet biogenesis does not appear to require coat proteins, whereas the phospholipid composition of the ER membrane would be crucial. In particular, phospholipids can influence the membrane surface tension, which is extremely important to the rounded shape purchase of LD. During this process, the neutral lipid area in contact with the aqueous media is reduced maximally and is determined by the phospholipid type dominance. For instance, while phospholipids, such as PE hinder biogenesis, lysophospholipids enhance it. These aspects also determine the gemmating direction, releasing lipid droplets into the cytosol, although they can be eventually released into the ER lumen [70–72].

Nonetheless, it has been shown that some ER membrane located proteins (storage-inducing transmembrane, FIT) regulate lipid droplet budding. More

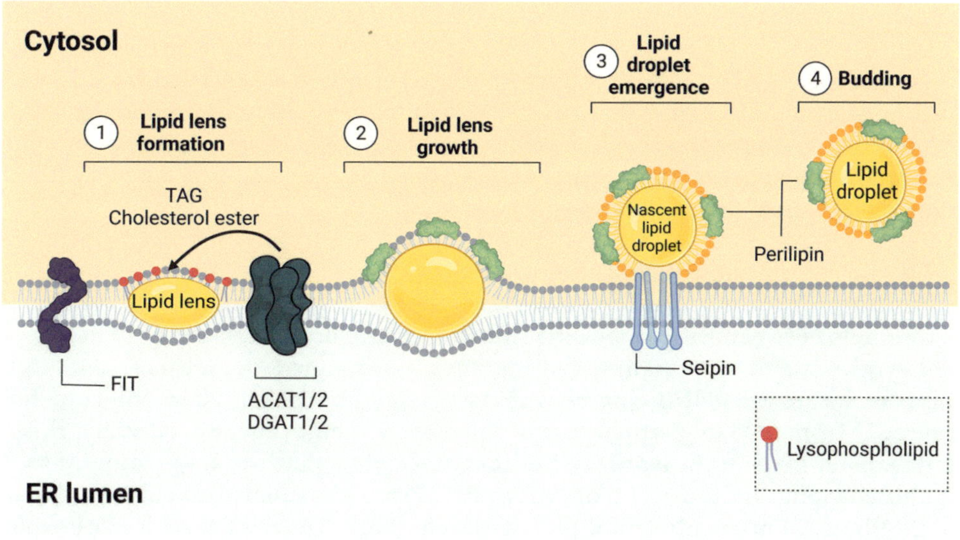


**Figure 4.**
*Graphical scheme of lipid droplet biogenesis from endoplasmic reticulum membrane. First, cholesterol esters and triglycerides (TAG) accumulate near the enzymes responsible for their synthesis; acyl-coenzymeA: Cholesterol acyltransferase 1/2 (ACAT1/2), and diacylglycerol acyltransferase 1/2 (DGAT1/2), respectively. This accumulation forms a lipid lens, a lipid droplet precursor. Lipid lens will grow, deforming ER membrane, a process promoted by certain phospholipids, such as lysophospholipids. To stabilize lipid droplet-ER contacts, Seipin is needed, helping in lipid droplet budding. Finally, when the lipid droplet formed has a correct size, it is released into the cytosol.*

specifically, FIT proteins maintain the ER lipidic composition and shape. In the absence of FIT, the quantity of sterol esters and triglycerides in the ER membrane increases through inhibition of lipid droplet gemmation. FIT seems to act by interacting directly with DAG; hence, a lack of FIT provokes DAG accumulation in ER, which inhibits the LD budding by altering ER morphology and membrane surface tension [73, 74].

In addition, the protein Seipin is also involved in the regulation of the formation of lipid droplets. This protein has important implications in stabilizing the contact sites between the ER and LD. Furthermore, when Seipin is absent, LD formation is delayed, meaning less incorporation of lipids and proteins, and so this leads to LDs generated becoming morphologically aberrant. Moreover, a protein related to peroxisomal biogenesis, Pex30, has been described to interact with Seipin in yeast. While in normal conditions Pex30 is located along the ER membrane, in Seipin mutants, it accumulates in LD biogenesis sites. When both Seipin and Pex30 are depleted, there is no LD biogenesis, neutral lipids are accumulated in ER membrane, peroxisomes are not well synthesized, and membranes present a phospholipid disbalance, increasing PC, phosphatidylinositol, and DAG. It is observed that Pex30 works in coordination with Seipin, controlling ER membrane lipid composition, especially in LD budding areas [75–77].

Once gemmated, lipid droplets will grow through different mechanisms that include lipid transfer from ER, LD fusion, or lipid synthesis in their membrane. Triglycerides can be synthesized in LD membranes due to the transfer of specific enzymes from ER. Moreover, ER supplies all the required phospholipids for the growth to LD [78, 79].

## 6. Endoplasmic reticulum and plasma membrane

The plasma membrane (PM) is a phospholipid bilayer that isolates the cell content from the outside. Although PM is mainly constituted of phospholipids (such as PC, PE, and PS), it also presents cholesterol, sphingolipids, and a huge variety of proteins, which allow the signal transduction. Furthermore, PM can establish contacts with ER where phospholipid and sterol transport occur [80, 81].

### 6.1 Plasma membrane-ER phospholipid transport

Contacts between ER and plasma membrane are mostly related to phospholipid synthesis and its signaling. For instance, during growth factors stimulation, cells translocate PITPNM1 (membrane-associated phosphatidylinositol transfer protein 1) from the Golgi apparatus to the plasma membrane. Specifically, this lipid transfer protein transports simultaneously phosphatidic acid from PM to ER and phosphatidylinositol from ER to PM. From phosphatidic acid and phosphatidylinositol many phospholipids with regulatory functions, such as phosphatidylinositol 4-phosphate (PI4P), DAG, or phosphatidylinositol 4,5-biphosphate ($PI4,5P_2$), can be synthesized and serve as precursors for the synthesis of PC, PS or PE. During high glucose concentration in pancreatic β-cells, it is observed another protein (phospholipid transfer protein C2CD2L, also known as TMEM24) with a similar role to PITPNM1 (it transports PI from ER to plasma membrane) [80, 82].

On the other hand, more proteins regulate lipid transport between ER and plasma membrane. For instance, ORP5/8 can transfer PS from ER to plasma membrane and PI4P in the opposite direction [80, 82]. All this continuous transport between these two organelles allows cells to main the main phospholipids levels and allows their signaling, which is essential in external and internal factors stimulation.

## 7. Main age-related alterations in ER and MCS

It was observed that in senescence cells, the endoplasmic reticulum increases its size, also altering its functionality. Moreover, there is variation in calcium concentration and their proteins, including chaperones, and glycosyltransferases. This alteration leads to a reduction of protein folding efficiency and an increase in misfolded protein accumulation, generating a long-term unfolded protein response. This permanent UPR activation forces cells to cell death and abnormal mitochondrial calcium signaling [83–85].

With age, alterations in MAMs, observed as an increase in the distance between ER and mitochondria, have also been seen, indicating its role in pathological conditions associated with age. This condition impairs calcium signaling and autophagy and increases ER stress. In fact, MAMs alterations have been reported in several age-related diseases, including cardiac cell senescence, cancer, inflammation, and metabolic diseases [83–85].

Apart from ER-mitochondria interactions, other types of relationships such as ER-plasma membrane, ER-Golgi, ER-lipid droplets, ER-lysosomes, ER-peroxisomes are interesting to explore in order to find out what occurs with aging.

## 8. Conclusions

The endoplasmic reticulum plays a central role in lipid homeostasis due to it establishing contact with essentially all cellular organelles, including mitochondria, peroxisomes, Golgi apparatus, lipid droplets, and plasma membrane (**Figure 5**).

To synthesize some of the more abundant glycerophospholipids, coordinated action between ER and mitochondria is needed, implicating several enzymes located in both organelles.

ER is also related to the biogenesis of peroxisomes and lipid droplets; peroxisomes interaction with ER can also generate cholesterol, essential in the plasma membrane fluidity maintenance, and plasmalogens, that protect cells from oxidative damage. As far as lipid droplets are concerned, they are generated from specific regions of the ER and their main function is to store triglycerides and cholesterol esters.

ER can also interact with the plasma membrane, exchanging phospholipids such as acid phosphatidic and phosphatidylinositol, leading to correct cellular signaling and response to extracellular stimuli. Finally, the endoplasmic reticulum also contacts the Golgi apparatus, an important event in sphingomyelin biosynthesis.

With age, it is documented that alteration in ER size and functionality, leading to a chronic UPR activation, induces apoptosis and aberrant calcium signaling. Moreover,

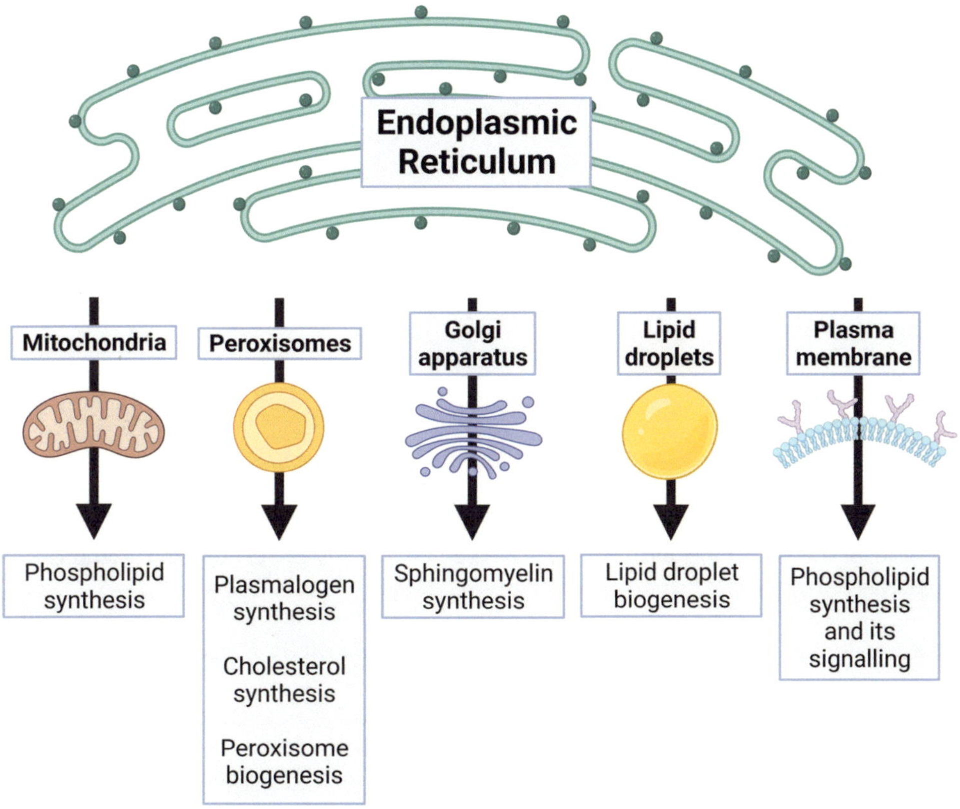


**Figure 5.**
*Scheme of endoplasmic reticulum contacts with multiple cellular organelles, such as mitochondria, peroxisomes, Golgi apparatus, lipid droplets, and plasma membrane; and, its functions.*

in aging models it was an increase in ER-mitochondrial distance was also observed, also altering the ER-mitochondria communication. However, the consequences of age in the other ER contacts nowadays are unclear.

Although we have explained some of the ER interactions, there is not enough information on all the synergistic functions that the ER has. Further research is needed.

## Acknowledgements

M. I. H-A acknowledges the support from the "Ayudas para contratos Ramón y Cajal (RYC) 2018" (RYC2018-024345-I) from the "Ministerio de Ciencia e Innovación" from Spain. We thank Inma Martínez-Ruiz and Alysha Jiwa for comments and the critical reading of this manuscript.

## Funding

This study was supported by research grants from MICINN (PID2019-105466RA-I00 AEI/ 10.13039/501100011033 and RYC2018-024345-I MCIN/AEI/ 10.13039/501100011033) and "la Caixa" Foundation project HR21-00430.

## Author details

Raúl Ventura[1] and María Isabel Hernández-Alvarez[1,2,3]*

1 Facultat de Biologia, Departament de Bioquímica i Biomedicina Molecular, Universitat de Barcelona, Barcelona, Spain

2 Centro de Investigación Biomédica en Red de Diabetes y Enfermedades Metabólicas Asociadas (CIBERDEM), Instituto de Salud Carlos III, Madrid, Spain

3 Institut de Biomedicina de la Universitat de Barcelona IBUB, Barcelona, Spain

*Address all correspondence to: mihernandez@ub.edu

## References

[1] Schwarz DS, Blower MD. The endoplasmic reticulum: Structure, function and response to cellular signaling. Cellular and Molecular Life Sciences. 2016;**73**:79-94. DOI: 10.1007/s00018-015-2052-6

[2] Westrate LM, Lee JE, Prinz WA, Voeltz GK. Form follows function: The importance of endoplasmic reticulum shape. Annual Review of Biochemistry. 2015;**84**:791-811. DOI: 10.1146/annurev-biochem-072711-163501

[3] Wu Y, Whiteus C, Xu CS, Hayworth KJ, Weinberg RJ, Hess HF, et al. Contacts between the endoplasmic reticulum and other membranes in neurons. Proceedings of the National Academy of Sciences of the United States of America. 2017;**114**:4859-4867. DOI: 10.1073/pnas.1701078114

[4] Mandon EC, Trueman SF, Gilmore R. Protein translocation across the rough endoplasmic reticulum. Cold Spring Harbor Perspectives in Biology. 2013;**5**: 1-14. DOI: 10.1101/cshperspect.a013342

[5] Lemos FO, Bultynck G, Parys JB. A comprehensive overview of the complex world of the endo- and sarcoplasmic reticulum Ca2+−leak channels. Biochimica et Biophysica Acta - Molecular Cell Research. 2021;**1868**:119020. DOI: 10.1016/j.bbamcr.2021.119020

[6] Zucker B, Kozlov MM. Mechanism of shaping membrane nanostructures of endoplasmic reticulum. Proceedings of the National Academy of Sciences of the United States of America. 2022;**119**:1-7. DOI: 10.1073/pnas.2116142119

[7] Neuhof A, Rolls MM, Jungnickel B, Kalies KU, Rapoport TA. Binding of signal recognition particle give ribosomes/nascent chain complexes a competitive advantages in endoplasmic reticulum membrane interaction. Molecular Biology of the Cell. 1998;**9**:103-115. DOI: 10.1091/mbc.9.1.103

[8] Zhao L, Cui Y, Fu G, Xu Z, Liao X, Zhang D. Signal recognition particle suppressor screening reveals the regulation of membrane protein targeting by the translation rate. MBio. 2021;**12**:1-19. DOI: 10.1128/mBio.02373-20

[9] Jomaa A, Eitzinger S, Zhu Z, Chandrasekar S, Kobayashi K, Shu-ou Shan NB. Molecular mechanism of cargo recognition and handover by the mammalian signal recognition particle. Cell Reports. 2021;**36**:109350. DOI: 10.1016/j.celrep.2021.109350

[10] Lee JH, Chandrasekar S, Chung SY, Fu YHH, Liu D, Weiss S, et al. Sequential activation of human signal recognition particle by the ribosome and signal sequence drives efficient protein targeting. Proceedings of the National Academy of Sciences of the United States of America. 2018;**115**:5487-5496. DOI: 10.1073/pnas.1802252115

[11] Furmanik M, Van Gorp R, Whitehead M, Ahmad S, Bordoloi J, Kapustin A, et al. Endoplasmic reticulum stress mediates vascular smooth muscle cell calcification via increased release of Grp78 (glucose-regulated protein, 78 kDa)-loaded extracellular vesicles. Arteriosclerosis, Thrombosis, and Vascular Biology. 2021;**78**:898-914. DOI: 10.1161/ATVBAHA.120.315506

[12] Tannous A, Pisoni GB, Hebert DN, Molinari M. N-linked sugar-regulated protein folding and quality control in the

ER. Seminars in Cell & Developmental Biology. 2015;**41**:79-89. DOI: 10.1016/j.semcdb.2014.12.001

[13] Lewy TG, Grabowski JM, Bloom ME. BiP: Master regulator of the unfolded protein response and crucial factor in flavivirus biology. The Yale Journal of Biology and Medicine. 2017;**90**(2):291-300 PMID: 28656015

[14] Jesus OM, Kim S, Sharp ME, JMJ S, Walther TC, Haven N, et al. Determinants of Endoplasmic Reticulum-to-Lipid Droplet Protein Targeting. Dev Cell. 2020;**54**(4):471-487. DOI: 10.1016/j.devcel.2020.07.001

[15] Hernández-Alvarez MI, Sebastián D, Vives S, Ivanova S, Bartoccioni P, Kakimoto P, et al. Deficient endoplasmic reticulum-mitochondrial phosphatidylserine transfer causes liver disease. Cell. 2019;**177**:881-895. DOI: 10.1016/j.cell.2019.04.010

[16] Knoblach B, Rachubinski RA. Determinants of the assembly, integrity and maintenance of the endoplasmic reticulum-peroxisome tether. Traffic. 2019;**20**:213-225. DOI: 10.1111/tra.12635

[17] Fox CA, Lethcoe K, Ryan RO. Calcium-induced release of cytochrome c from cardiolipin nanodisks: Implications for apoptosis. Biochimica et Biophysica Acta - Biomembranes. 2021;**1863**:183722. DOI: 10.1016/j.bbamem.2021.183722

[18] Kalpage HA, Wan J, Morse PT, Zurek MP, Turner AA, Khobeir A, et al. Cytochrome c phosphorylation: Control of mitochondrial electron transport chain flux and apoptosis. The International Journal of Biochemistry & Cell Biology. 2020;**121**:105704. DOI: 10.1016/j.biocel.2020.105704

[19] Sharma P, Sampath H. Mitochondrial DNA integrity: Role in health and disease. Cell. 2019;**8**:100. DOI: 10.3390/cells8020100

[20] Chang TC, Lee HT, Pan SC, Cho SH, Cheng C, Ou LH, et al. Metabolic reprogramming in response to alterations of mitochondrial DNA and mitochondrial dysfunction in gastric adenocarcinoma. International Journal of Molecular Sciences. 2022;**23**:1857. DOI: 10.3390/ijms23031857

[21] Yan C, Duanmu X, Zeng L, Liu B, Song Z. Mitochondrial DNA: Distribution, mutations, and elimination. Cells. 2019;**1**:1-15. DOI: 10.3390/cells8040379

[22] Yeo HK, Park TH, Kim HY, Jang H, Lee J, Hwang G, et al. Phospholipid transfer function of PTPIP51 at mitochondria-associated ER membranes. EMBO Reports. 2021;**22**:1-14. DOI: 10.15252/embr.202051323

[23] Manganelli V, Matarrese P, Antonioli M, Gambardella L, Vescovo T, Gretzmeier C, et al. Raft-like lipid microdomains drive autophagy initiation via AMBRA1-ERLIN1 molecular association within MAMs. Autophagy. 2021;**17**:2528-2548. DOI: 10.1080/15548627.2020.1834207

[24] Moser von Filseck J, Čopič A, Delfosse V, Vanni S, Jackson CL, Bourguet W, et al. Phosphatidylserine transport by ORP/Osh proteins is driven by phosphatidylinositol 4-phosphate. Science. 2015;**349**:432-436. DOI: 10.1126/science.aab1346

[25] Morciano G, Marchi S, Morganti C, Sbano L, Bittremieux M, Kerkhofs M, et al. Role of mitochondria-associated ER membranes in calcium regulation in Cancer-specific settings. Neoplasia. 2018;**20**:510-523. DOI: 10.1016/j.neo.2018.03.005

[26] Senoo N, Miyoshi N, Kobayashi E, Morita A, Tanihata J, Takeda S, et al. Glycerophospholipid profile alterations are associated with murine muscle-wasting phenotype. Muscle & Nerve. 2020;**62**:413-418. DOI: 10.1002/mus.26993

[27] Annibal A, Riemer T. Europe PMC Funders Group Structural, biological and biophysical properties of glycated and glycoxidized phosphatidylethanolamines. 2018;**49**:293-307. DOI: 10.1016/j.freeradbiomed.2016.03.011

[28] Walkey CJ, Yu L, Agellon LB, Vance DE. Biochemical and evolutionary significance of phospholipid methylation. The Journal of Biological Chemistry. 1998;**273**:27043-27046. DOI: 10.1074/jbc.273.42.27043

[29] Quiroga IY, Pellon-Maison M, Suchanek AL, Coleman RA, Gonzalez-Baro MR. Glycerol-3-phosphate acyltransferases 3 and 4 direct glycerolipid synthesis and affect functionality in activated macrophages. The Biochemical Journal. 2019;**476**:85-99. DOI: 10.1042/BCJ20180381

[30] Vance JE. Phospholipid synthesis and transport in mammalian cells. Traffic. 2015;**16**:1-18. DOI: 10.1111/tra.12230

[31] Borkenhagen LF, Kennedy EP, Fielding L. Enzymatic formation and decarboxylation of phosphatidylserine. The Journal of Biological Chemistry. 1961;**236**:28-30. DOI: 10.1016/S0021-9258(19)63319-3

[32] Kennedy EP, Weiss SB. The function of cytidine coenzymes in the biosynthesis of phospholipides. The Journal of Biological Chemistry. 1956;**222**:193-214 PMID: 13366993

[33] Wan S, van der Veen JN, N'Goma JCB, Nelson RC, Vance DE, Jacobs RL. Hepatic PEMT activity mediates liver health, weight gain, and insulin resistance. The FASEB Journal. 2019;**33**:10986-10995. DOI: 10.1096/fj.201900679R

[34] Miyata N, Kuge O. Topology of phosphatidylserine synthase 1 in the endoplasmic reticulum membrane. Protein Science. 2021;**30**(11):2346-2353. DOI: 10.1002/pro.4182

[35] Kannan M, Lahiri S, Liu LK, Choudhary V, Prinz WA. Phosphatidylserine synthesis at membrane contact sites promotes its transport out of the ER. Journal of Lipid Research. 2017;**58**:553-562. DOI: 10.1194/jlr.M072959

[36] Chen L, Liu B, Qin Y, Li A, Gao M, Liu H, et al. Mitochondrial fusion protein Mfn2 and its role in heart failure. Frontiers in Molecular Biosciences. 2021;**8**:1-10. DOI: 10.3389/fmolb.2021.681237

[37] Watanabe Y, Tamura Y, Kakuta C, Watanabe S, Endo T. Structural basis for interorganelle phospholipid transport mediated by VAT-1. The Journal of Biological Chemistry. 2020;**295**:3257-3268. DOI: 10.1074/jbc.RA119.011019

[38] Zhong W. Oxysterol-binding protein-related protein 8 (ORP8) increases sensitivity of hepatocellular carcinoma cells to Fas-mediated apoptosis.Journal of Biological Chemistry 2015;**290**:12. DOI: 10.1074/jbc.M114.610188

[39] Okumoto K, Tamura S, Honsho M, Fujiki Y. Peroxisome: Metabolic functions and biogenesis. Advances in Experimental Medicine and Biology. 2020;**1299**:3-17. DOI: 10.1007/978-3-030-60204-8_1

[40] Fransen M, Lismont C, Walton P. The peroxisome-mitochondria connection: How and why? International Journal of Molecular Sciences. 2017;**18**:1126. DOI: 10.3390/ijms18061126

[41] Wanders RJA, Waterham HR, Ferdinandusse S. Metabolic interplay between peroxisomes and other subcellular organelles including mitochondria and the endoplasmic reticulum. Frontiers in Cell and Development Biology. 2016;**3**:1-15. DOI: 10.3389/fcell.2015.00083

[42] Nitta Y, Muraoka-Hirayama S, Sakurai K. Catalase is required for peroxisome maintenance during adipogenesis. Biochimica et Biophysica Acta - Molecular and Cell Biology of Lipids. 2020;**1865**:158726. DOI: 10.1016/j.bbalip.2020.158726

[43] Park H, He A, Tan M, Johnson JM, Dean JM, Pietka TA, et al. Peroxisome-derived lipids regulate adipose thermogenesis by mediating cold-induced mitochondrial fission. The Journal of Clinical Investigation. 2019;**129**:694-711. DOI: 10.1172/JCI120606

[44] Horlenko O, Lenchenko A, Pushkarenko O, Kossey G, Tomey A. Impairment of peroxisome biogénesis in the Spectrum of Zellweger syndrome (clinical case). Georgian Medical News. 2021;(318):60-67. PMID: 34628380

[45] Jansen RLM, van der Klei IJ. The peroxisome biogenesis factors Pex3 and Pex19: Multitasking proteins with disputed functions. FEBS Letters. 2019;**593**:457-474. DOI: 10.1002/1873-3468.13340

[46] Dimitrov L, Lam SK, Schekman R. The role of the endoplasmic reticulum in peroxisome biogenesis. Cold Spring Harbor Perspectives in Biology. 2013;**5**:1-12. DOI: 10.1101/cshperspect.a013243

[47] Charles KN, Shackelford JE, Faust PL, Fliesler SJ, Stangl H, Kovacs WJ. Functional peroxisomes are essential for efficient cholesterol sensing and synthesis. Frontiers in Cell and Development Biology. 2020;**8**:560266. DOI: 10.3389/fcell.2020.560266

[48] Honsho M, Fujiki Y. Plasmalogen homeostasis – Regulation of plasmalogen biosynthesis and its physiological consequence in mammals. FEBS Letters. 2017;**591**:2720-2729. DOI: 10.1002/1873-3468.12743

[49] Cui W, Liu D, Gu W, Chu B. Peroxisome-driven ether-linked phospholipids biosynthesis is essential for ferroptosis. Cell Death and Differentiation. 2021;**28**:2536-2551. DOI: 10.1038/s41418-021-00769-0

[50] Messias MCF, Mecatti GC, Priolli DG, De Oliveira CP. Plasmalogen lipids: Functional mechanism and their involvement in gastrointestinal cancer. Lipids in Health and Disease. 2018;**17**:1-12. DOI: 10.1186/s12944-018-0685-9

[51] Xiao J, Luo J, Hu A, Xiao T, Li M, Kong Z, et al. Cholesterol transport through the peroxisome-ER membrane contacts tethered by PI(4,5)P2 and extended synaptotagmins. Science China. Life Sciences. 2019;**62**:1117-1135. DOI: 10.1007/s11427-019-9569-9

[52] Kovacs WJ, Tape KN, Shackelford JE, Duan X, Kasumov T, Kelleher JK, et al. Localization of the pre-squalene segment of the isoprenoid biosynthetic pathway in mammalian peroxisomes. Histochemistry and Cell Biology. 2007;**127**(3):273-290. DOI: 10.1007/s00418-006-0254-6

[53] Liu J, Huang Y, Li T, Jiang Z, Zeng L, Hu Z. The role of the Golgi apparatus in disease (review). International Journal of Molecular Medicine. 2021;**47**:38. DOI: 10.3892/ijmm.2021.4871

[54] Li J, Ahat E, Wang Y. Golgi structure and function in health, stress, and

diseases. Results and Problems in Cell Differentiation. 2019;**67**:441-485. DOI: 10.1007/978-3-030-23173-6_19

[55] Frappaolo A, Karimpour-Ghahnavieh A, Sechi S, Giansanti MG. The close relationship between the Golgi trafficking machinery and protein glycosylation. Cell. 2020;**9**(12):2652. DOI: 10.3390/cells9122652

[56] Toshima K, Nagafuku M, OkazakiT, KobayashiT, InokuchiJI. Plasma membrane sphingomyelin modulates thymocyte development by inhibiting TCR-induced apoptosis. International Immunology. 2019;**31**:211-223. DOI: 10.1093/intimm/dxy082

[57] Gault CR, Obeid LM, Hannun YA. An overview of sphingolipid metabolism: From synthesis to breakdown. Advances in Experimental Medicine and Biology. 2010;**688**:1-23. DOI: 10.1007/978-1-4419-6741-1_1

[58] Kumagai K, Hanada K. Structure, functions and regulation of CERT, a lipid-transfer protein for the delivery of ceramide at the ER–Golgi membrane contact sites. FEBS Letters. 2019;**593**:2366-2377. DOI: 10.1002/1873-3468.13511

[59] Fujimoto T, Parton RG. Not just fat: The structure and function of the lipid droplet. Cold Spring Harbor Perspectives in Biology. 2011;**3**:1-17. DOI: 10.1101/cshperspect.a004838

[60] Jackson CL. Lipid droplet biogenesis. Current Opinion in Cell Biology. 2019;**59**:88-96. DOI: 10.1016/j.ceb.2019.03.018

[61] Benador IY, Veliova M, Mahdaviani K, Petcherski A, Jakob D, Assali E, et al. Mitochondria bound to lipid droplets have unique bioenergetics, composition, and dynamics that support lipid droplet expansion. Cell Metabolism. 2018;**27**(4):869-885. DOI: 10.1016/j.cmet.2018.03.003

[62] Barisch C, Paschke P, Hagedorn M, Maniak M, Soldati T. Lipid droplet dynamics at early stages of Mycobacterium marinum infection in Dictyostelium. Cellular Microbiology. 2015;**17**:1332-1349. DOI: 10.1111/cmi.12437

[63] Tsai TH, Chen E, Li L, Saha P, Lee HJ, Huang LS, et al. The constitutive lipid droplet protein PLIN2 regulates autophagy in liver. Autophagy. 2017;**13**:1130-1144. DOI: 10.1080/15548627.2017.1319544

[64] Nguyen TB, Louie SM, Daniele JR, Tran Q, Dillin A, Zoncu R, et al. DGAT1-dependent lipid droplet biogenesis protects mitochondrial function during starvation-induced autophagy. Developmental Cell. 2017;**14**:535-562. DOI: 10.1016/j.devcel.2017.06.003

[65] Voisin P, Bernard M, Bergès T, Régnacq M. Amino acid starvation inhibits autophagy in lipid droplet-deficient cells through mitochondrial dysfunction. The Biochemical Journal. 2020;**477**:3613-3623. DOI: 10.1042/BCJ20200551

[66] Gaspar ML, Hofbauer HF, Kohlwein SD, Henry SA. Coordination of storage lipid synthesis and membrane biogenesis: Evidence for cross-talk between triacylglycerol metabolism and phosphatidylinositol synthesis. The Journal of Biological Chemistry. 2011;**286**:1696-1708. DOI: 10.1074/jbc.M110.172296

[67] Khandelia H, Duelund L, Pakkanen KI, Ipsen JH. Triglyceride blisters in lipid bilayers: Implications for lipid droplet biogenesis and the mobile lipid signal in cancer cell membranes. PLoS One. 2010;**5**:1-8. DOI: 10.1371/journal.pone.0012811

[68] Duelund L, Jensen GV, Hannibal-Bach HK, Ejsing CS, Pedersen JS, Pakkanen KI, et al. Composition, structure and properties of POPC-triolein mixtures. Evidence of triglyceride domains in phospholipid bilayers. Biochimica et Biophysica Acta - Biomembranes. 2013;**1828**:1909-1917. DOI: 10.1016/j.bbamem.2013.03.020

[69] Choudhary V, Ojha N, Golden A, Prinz WA. A conserved family of proteins facilitates nascent lipid droplet budding from the ER. The Journal of Cell Biology. 2015;**211**:261-271. DOI: 10.1083/jcb.201505067

[70] Chorlay A, Thiam AR. An asymmetry in monolayer tension regulates lipid droplet budding direction. Biophysical Journal. 2018;**114**:631-640. DOI: 10.1016/j.bpj.2017.12.014

[71] Ben M'barek K, Ajjaji D, Chorlay A, Vanni S, Forêt L, Thiam AR. ER membrane phospholipids and surface tension control cellular lipid droplet formation. Developmental Cell. 2017;**41**:591-604. DOI: 10.1016/j.devcel.2017.05.012

[72] Choudhary V, Golani G, Joshi A, Cottier S, Schneiter R, Prinz W. Architecture of lipid droplets in endoplasmic reticulum is determined by phospholipid intrinsic curvature. Current Biology. 2019;**28**(6):915-926. DOI: 10.1016/j.cub.2018.02.020

[73] Hayes MJ, Choudhary V, Ojha N, Shin JJH, Han GS, Carman GM, et al. Fat storage-inducing transmembrane (FIT or FITM) proteins are related to lipid phosphatase/ phosphotransferase enzymes. Microbial Cell. 2018;**5**:88-103. DOI: 10.15698/mic2018.02.614

[74] Becuwe M, Bond LM, Pinto AFM, Boland S, Mejhert N, Elliott SD, Cicconet M, Graham MM, Liu XN, Ilkayeva O, Saghatelian A, Walther TC, Farese RV Jr. FIT2 is an acyl-coenzyme A diphosphatase crucial for endoplasmic reticulum homeostasis. Journal of Cell Biology. 2020;**219**(10):202006111. DOI: 10.1083/jcb.202006111

[75] Arlt H, Sui X, Folger B, Adams C, Chen X, Remme R, et al. Seipin forms a flexible cage at lipid droplet formation sites. Nature Structural & Molecular Biology. 2022;**29**:194-202. DOI: 10.1038/s41594-021-00718-y

[76] Joshi AS, Nebenfuehr B, Choudhary V, Satpute-Krishnan P, Levine TP, Golden A, et al. Lipid droplet and peroxisome biogenesis occur at the same ER subdomains. Nature Communications. 2018;**9**:1-12. DOI: 10.1038/s41467-018-05277-3

[77] Wang S, Idrissi FZ, Hermansson M, Grippa A, Ejsing CS, Carvalho P. Seipin and the membrane-shaping protein Pex30 cooperate in organelle budding from the endoplasmic reticulum. Nature Communications. 2018;**9**:1-12. DOI: 10.1038/s41467-018-05278-2

[78] Krahmer N, Guo Y, Wilfling F, Hilger M, Lingrell S, Heger K, et al. Phosphatidylcholine synthesis for lipid droplet expansion is mediated by localized action of CTP: Phosphocholine Cytidylyltransferase. Cell Metabolism. 2011;**14**:504-515. DOI: 10.1016/j.cmet.2011.07.013

[79] Barneda D, Christian M. Lipid droplet growth: Regulation of a dynamic organelle. Current Opinion in Cell Biology. 2017;**47**:9-15. DOI: 10.1016/j.ceb.2017.02.002

[80] Beh CT. Sticking with it: ER-PM membrane contact sites as a coordinating Nexus for regulating lipids and proteins at the cell cortex. Frontiers in Cell and Development Biology. 2020;**8**:23. DOI: 10.3389/fcell.2020.00675

[81] Le Roux AL, Quiroga X, Walani N, Arroyo M, Roca-Cusachs P. The plasma membrane as a mechanochemical transducer. Philosophical Transactions of the Royal Society B. 2019;**374**:20180221. DOI: 10.1098/rstb.2018.0221

[82] Kim YJ, Guzman-Hernandez ML, Wisniewski E, Balla T. Phosphatidy linositol-phosphatidic acid exchange by Nir2 at ER- PM contact sites maintains phosphoinositide Signaling competence. Dev Cell. 2016;**25**:549-561. DOI: 10.1016/j.devcel.2015.04.028

[83] Ruiz-Meana M, Minguet M, Bou-Teen D, Miro-Casas E, Castans C, Castellano J, et al. Ryanodine receptor glycation Favors mitochondrial damage in the senescent heart. Circulation. 2019;**12**(139):949-964. DOI: 10.1161/CIRCULATIONAHA.118.035869

[84] Tubbs E, Rieusset J. Metabolic signaling functions of ER–mitochondria contact sites: Role in metabolic diseases. Journal of Molecular Endocrinology. 2017;**58**:R87-R106. DOI: 10.1530/JME-16-0189

[85] Csordás G, Weaver D, Hajnóczky G. Endoplasmic reticulum–mitochondrial Contactology: Structure and Signaling functions. Trends in Cell Biology. 2018;**28**:523-540. DOI: 10.1016/j.tcb.2018.02.009

Chapter 3

# Responses of Endoplasmic Reticulum to Plant Stress

*Vishwa Jyoti Baruah, Bhaswati Sarmah, Manny Saluja and Elizabeth H. Mahood*

## Abstract

Global climate change has resulted in alterations in the biotic and abiotic conditions of the planet. This has led to changes in the agricultural system resulting from reduced water availability, increased temperature increase in the population and occurrences of pests and diseases. Plants are adversely affected when they experience any stress retarding their growth, development and productivity. Endoplasmic Reticulum (ER) is an organelle that shows a tremendous response when subjected to stress conditions. Therefore, to explore and comprehend plants' multidimensional interactions when subjected to stress conditions, an insight into the molecular stress signalling in the ER in response to the stress situation is discussed in this chapter.

**Keywords:** biotic stress, cold stress, drought, endoplasmic reticulum, heat stress, salt stress, plant defence

## 1. Introduction

The endoplasmic reticulum (ER) is a versatile, dynamic and largely pleiotropic subcellular organelle forming an essential part of eukaryotes. It is one of the largest in size, complex in functionality and variable in architecture [1]. It plays a significant role in maintaining the spatial organisation of endomembrane organelles by acting as an architectural framework. It also synthesises essential cellular building blocks like lipids and proteins. ER lies at the core of the endomembrane system, which comprises unified endocytic and biosynthetic cellular processes and is composed of two different structural subdomains. One is the nuclear envelope enclosing the nucleus, and the other is the peripheral ER comprising the interconnected network of flattened sacs and tubules [2, 3]. At a submicron level, the ER network is organised in morphologically distinct domains that assume specific functions [4]. The ER represents the organelle with the largest membrane surface area owing to its network of interconnected tubules and flattened cisternae. Several genetic studies aided with live-cell imaging have illustrated the underlying drivers and the implausible dynamism of ER. The ER exemplifies a secretory pathway gatekeeper that controls protein quality control, its folding, signalling, and degradation across multiple checkpoints and impacts the general plant growth cellular homeostasis.

The ER is responsible for synthesising an array of proteins such as enzymes, ion channels, receptors and cargo molecules that modulate several essential physiological processes, which are ultimately either retained in or distributed from the ER [5–7]. Besides syntheses, ER also serves as a storehouse for carbohydrates and calcium [8, 9] and plays an essential role in abiotic and biotic stress resistance through the control of protein folding and signalling [10]. These characteristics of ER make it a functionally and structurally non-uniform yet morphologically continuous cellular compartment.

## 2. ER in plants: structure and function

In plant cells, the ER plays a critical role in the organism and bears a strong connection with other plant organelles like the vacuole, Golgi apparatus [7, 11, 12] and chloroplast [13, 14] forming the shape of a spider-webbed membrane network and any defect in its functionality can give rise to several developmental defects [15–18]. Based on electron microscopy studies, the ER can be differentiated into smooth ER-bearing subdomains with associated ribosomes, rough ER with subdomains of ribosomes-free regions, and nuclear envelope regions which are enwrapped by the ER forming a double membrane demarcating the nucleus [4, 19, 20]. The ER forms connections with the neighbouring cells through channels called plasmodesmata that protrude into the cell wall and interconnect with the cytoplasm of the neighbouring cells [21, 22], forming a unique organelle that cannot be delimited by a cell boundary. Electron microscopy-aided studies have also reported that the plant ER form connections with other membranes, including vacuoles, plastids, mitochondria and Golgi apparatus. In addition to the ER- plasma membrane contact sites (EPCSs) and plasmodesmata [4, 23]. The optical trapping and tweezer system has established physical contact of the chloroplast and Golgi apparatus with the ER [24, 25].

It has been thought that the movement of the ER influences the movement of other organelles, which is possible by the several network connections from the ER to the other cell organelles. While the ER enlarges, contracts or reorganises its morphology, its integrity is maintained by unfolded protein response (UPR) [26] signalling demonstrating the relationship between the ER morphology and its functional integrity [27–29]. Therefore, the connection of ER with other cellular organelles and compartments is crucial for the exchange of constituents as well as for their function and spatial distribution [29, 30].

The most fascinating feature of the ER in the plant cell is its extraordinary dynamicity which can be easily observed through microscopic analysis resulting in the overall evolving architecture of the tubules and cisternae of the ER that are in a state of continuous movement and rearrangement [27, 29, 31]. This morphological reorganisation of the ER has been reported previously where in the initial phases of cell growth, the ER was observed as a compact entanglement of membranes which undergo reorganisation into an open network as the cell growth advances [27, 32]. The ER organisations become even more complicated in root cells where the ER assumes a condensed form, primarily where the root hairs originate [27, 33]. Evidence of a link between ER shape and ER function can be demonstrated when mutations in Root Hair Defective3 (RHD3) ER-shaping protein compromise the functional organisation of the ER by elongation of the cells and transitioning to a reticulate pattern from a more sheet-like form [29, 30].

## 3. Response of plant ER to stress

The changing climate poses a significant risk to the plants owing to the transitions in its natural environment from typical to very harsh conditions. When plants are exposed to adversities like low water availability, high or chilling temperature, salt concentrations in the soil or pest and disease infestations, they usually respond by changing their response in a dynamic way. Although several studies have aimed to investigate the response of plants to individual environmental stress, in the field, the scenario is such that the plants may be exposed to multiple stresses, and their response may be quite different from that in the laboratory [34]. These dynamic responses in plants include changing the proteome by changing the gene expression patterns [35]. In plants, the ER is the prime organelle that regulates the stress responses [36, 37], as any stress which affects protein folding leads to a cellular homeostatic response mediated by the UPR in response to ER stress [36, 38]. ER is the point of synthesis of the majority of plant cell proteins and also the point of folding of unfolded or misfolded proteins, which are aided by chaperons and co-chaperons [39, 40]. The ER quality control (ERQC) and ER-associated degradations (ERAD) are two mechanisms that maintain the folding of proteins and degrade the misfolded proteins, respectively [39]. Nevertheless, sometimes, these mechanisms are surpassed by the UPR when there is an accumulation and persistence of misfolded and/or unfolded proteins leading to a state where the organelles trigger specific signalling pathways to restore the ER mechanism and recover from the stress [41].

## 4. ER stress signalling in response to heat stress

Heat stress often poses a serious threat to plants, as it negatively affects photosynthesis and fertility and can at times result in death. At the cellular level, symptoms of heat stress include the accumulation of misfolded or denatured proteins, production of ROS, microtubule disorganisation, and membrane instability. As the endoplasmic reticulum (ER) is the site of production for many proteins and lipids, this organelle is tightly linked with the heat stress response. While heat stress response mechanisms include the accumulation of osmolytes and antioxidants, perhaps the most canonical response is the production of Heat Shock Proteins (HSPs): molecular chaperones which facilitate proper protein folding and/or disaggregation of misfolded proteins [42].

Two critical components of the ER's Unfolded Protein Response (UPR) are also the major regulators of the heat stress response: the transcription factor bZIP28 and cytosolic RNA-splicing enzyme IRE1. bZIP28 was first determined to be involved in the heat stress response in *Arabidopsis* through a combination of co-expression analysis, cellular localisation studies, and reverse genetics [43]. Further studies have elucidated the mechanism of action of bZIP28 in the heat stress response [44, 45]. Briefly, in unstressed cells, bZIP28 is tethered to the ER membrane by an HSP (Binding Immunoglobulin Protein, or BiP), but when misfolded or denatured proteins accumulate in the ER, BIP is competed away from bZIP28, thereby releasing it to act as a transcription factor and upregulate the expression of other HSPs. *Arabidopsis* IRE1 was found to upregulate bZIP60 – another transcription factor upregulating stress-responsive genes – through cytosolic mRNA splicing [46]. These proteins are at once essential for the UPR stress response in general and the heat stress response, specifically as HSPs are among their direct or indirect targets and their loss of function results in decreased survival under heat stress [47]. These initial studies uncovered the ER's direct role in mediating heat stress.

More recent studies have discovered roles for the ER in the metabolism under heat stress, specifically in the starch and lipid biosynthesis pathways. Although the role of the ER in starch metabolism is less clear than its role in lipid metabolism, several studies have implicated its involvement. In rice, heat stress during grain set can result in a "chalky" grain appearance and decreased starch content. Several studies have found altered expression of starch biosynthetic genes in mutants of the UPR, particularly in seeds [48, 49]. Other studies have found that the application of heat stress during the pre-cellularisation stage of grain set results in increased chalkiness and decreased starch levels [50]. A final study found that components of the UPR regulate starch content and chalkiness in rice seeds under heat stress [51]. When considered together, these results implicate that the ER/UPR regulates starch metabolism under heat stress. While the exact mechanisms underpinning this regulation are unclear, it has been known for some time that the ER is a site of lipid biosynthesis in plants. Tight control of lipid metabolism under heat stress is required as plants alter the lipid composition of membranes in order to combat heat-induced membrane instability.

In many plants, including *Arabidopsis* [52], tomato [53], wheat [54], and turfgrasses [55], an increase in the amounts of ER-produced lipid classes occurs under heat. Forward genetic studies have further demonstrated that ER stress, as well as heat stress, can alter lipid metabolism. For example, *Arabidopsis* mutants with inactive Fatty Acid Desaturase 2 (fad2 mutants) show either aberrant phenotypes or increased symptoms under heat [56] or ER stress (imposed by tunicamycin) [57], respectively, compared to WT. In soybean, isoforms of FAD2 were found to be unstable in high temperatures [58], and FAD2 transcripts were found to be downregulated under heat stress in *Arabidopsis* [59]. These results suggest that lipid metabolism, and to some degree starch metabolism, is dependent upon favourable conditions in the ER and is altered under heat stress.

One of the open questions in ER stress regulation is how the stress is perceived by the ER and what molecules or phenomena serve as the stress signal(s). Extracellular $Ca^{2+}$ was one of the first candidates hypothesised to be the cellular signal for heat stress [60]. Seminal studies hypothesised the following signalling response: the influx of extracellular $Ca^{2+}$ via Cyclic Nucleotide Gated Channels activates Calmodulin, which activates kinases, which finally activates Heat Shock Transcription Factors [61]. Other early putative signals included a nuclear-localised histone sensor and two ER-localised unfolded protein sensors [61]. More recent studies have implicated other molecules to act as a signal for heat stress, specifically Jasmonic Acid (JA) and phytochromes. In a recent study in rice grains, JA biosynthesis and signalling activity was enriched among genes upregulated one hour after heat shock, and JA accumulated three hours after heat. Furthermore, treatment with Methyl JA (MeJA) increased the abundance of IRE1-spliced OsBZIP50(s), an ER-stress response biomarker [50]. In another study conducted in *Arabidopsis* seeds, HY5, a component of light signalling that is downstream of phytochromes, was found to negatively regulate the UPR [62]. Taken concurrently with evidence that phytochromes have thermo-sensing capabilities and that null phytochrome mutants exhibit a constitutive heat stress phenotype [63], the above study provides evidence that phytochromes may sense and transmit the heat stress signal to the ER. Further research is needed to determine the conditions under which these signals are active and if they act in conjunction with each other.

Heat stress can be particularly damaging to anthers and to the developing pollen. Recent evidence points to the UPR being constitutively active in male reproductive tissues [64], and therefore it is critical to understand how the UPR is affected by heat stress in these tissues. In a recent study, several ER-stress genes, including BiP, other chaperones/HSPs, and genes involved in the ER protein cleaning system, were all upregulated under heat stress in pollen germinating *in vitro* [65]. Furthermore, through forward-genetic studies, it was recently discovered that ER-localised chaperones are critical for proper pollen development and seed set under heat stress. *Arabidopsis* mutants of the Thermosensitive Male Sterile 1 (TMS1) gene, encoding an HSP40, showed reduced fertility and pollen coat abnormalities when grown under 29°C [66]. Similarly, *Arabidopsis* mutants lacking IRE1 were male sterile at 29°C yet had viable pollen when grown at room temperature [64]. As proper protein production and secretion are critical for pollen tube growth, any disruption to this process – such as that caused by heat stress – is hypothesised to be detrimental to pollen growth and viability and, therefore, negatively affect fertility.

## 5. ER stress signalling in response to drought stress

Drought stress is a major driver of yield losses – with a tremendous potential to limit plant growth than any other abiotic stress [67]. An important drought response mechanism is the closure of stomata. This at once conserves water by decreasing evapotranspiration and lowers photosynthesis rates, leading to decreased growth rates. If a period of drought occurs during the reproductive or grain filling stage, yield losses can occur due to pollen sterility or embryo abortion. Plant responses and adaptions to drought stress include the aforementioned stomatal closure, deeper root growth, reduced leaf size and altered leaf orientation. At the cellular level, these drought responses result in loss of turgor, which stimulates the production of the hormone Abscisic acid (ABA). ABA is a canonical "stress response" hormone [68], bringing about morphological changes through the following signal transduction cascade: ABA molecules bind to their receptors, which causes inhibition of PP2C phosphatases, leading to the autophosphorylation and activation of SUCROSE NONFERMENTING1-RELATED SUBFAMILY 2 (SnRK2) kinases, which phosphorylate ABA-responsive transcription factors (AREBs), ultimately leading to the expression of stress-responsive genes.

Some abiotic stresses, notably salt and heat stress, stimulate ER stress by markedly increasing the numbers of misfolded proteins in a cell. However, the mechanism by which drought stress is related to ER stress has yet to be determined. Toward this end, several studies have found multiple components of the ER protein folding machinery to be essential for proper drought response, including several E3 ubiquitin ligases, the transcription factor bZIP60, and Binding Protein (BiP). BiP is a molecular chaperone that contributes to proper protein folding, especially under ER stress. Overexpression of BiP has been found to confer drought tolerance - specifically less stomatal closure, less wilting, and maintenance of turgor - in *Nicotiana tabacum* (tobacco) and *Glycine max* (soybean) [69, 70]. Interestingly, the drought tolerance was concurrent with a decrease in the levels of typical drought stress response mechanisms, including the osmolytes proline/sucrose/glucose, root biomass, and drought stress response genes NAC2, glutathione-S-transferase, and antiquitin [69]. Similar to BiP, E3 ligases play essential roles in ER protein folding. The putative E3 ubiquitin ligase SUPPRESSOR OF DRY2 DEFECTS1 (SUD1) is homologous to the mammalian TEB4, a component of the ER-associated degradation pathway (ERAD) that marks non-functional proteins for destruction [71]. Both TEB4 and *Arabidopsis* SUD1 are implicated in the regulation of sterol biosynthesis [71, 72] – an interesting link between the ER and cellular stress response as sterols are components of lipid membranes that can change membrane fluidity as the temperature fluctuates. Notably, mutations in SUD1 were found to restore drought hypersensitive2 (dry2) [72] – which show increased sensitivity to drought stress due to abnormal sterol composition in roots and aberrant ROS signalling [73] – to WT phenotypes, revealing a link between the ERAD and drought responses. Two additional E3 ubiquitin ligases, Rma1H1 (from *Capsicum annum*) and Rma1 (from *Arabidopsis*), confer drought tolerance when overexpressed, putatively through suppressing the trafficking of the aquaporin channel PIP2;1 [74]. The transcription factor bZIP60 – the target of the mRNA splicing enzyme IRE1 – is a third component of the ER stress response that confers drought tolerance when overexpressed. bZIP60 from *Boea hygrometrica* (the extremely drought-tolerant "resurrection plant") conferred drought tolerance to *Arabidopsis* when overexpressed through the up-regulation of several ER quality control genes [75]. These results, when considered together, suggest that increased activity of the ERAD provides enhanced drought tolerance and therefore that ER functions are essential for survival under drought stress, though further research is needed to pinpoint the specific functions of the ER components under drought stress.

Several studies have uncovered an interesting link between ER stress and ABA signalling under drought stress – suggesting that ER functions are mechanistically related to drought stress through ABA. One study mentioned above transformed bZIP60 from *B. hygrometrica* (BhbZIP60) into *Arabidopsis*, overexpressed it, and found enhanced drought tolerance and increased expression of both ER components and ABA-responsive genes [75]. Interestingly, the authors found that BhbZIP60 was able to bind to the ERSE cis-regulatory elements present in ER stress-responsive genes but unable to bind to ABA-responsive cis-regulatory elements. Further, BhbZIP60 (in its native host *B. hygrometrica*), was found to be highly upregulated under drought but not so under heat or salt – which highly upregulates bZIP60 expression in *Arabidopsis* [46]. Another study found that inactivating the ERSE in *Zea mays* PP2C-A gene – which encodes a phosphatase with important roles in ABA signal transduction – confers drought stress tolerance [76]. While the WT (ERSE inactive) promotor was responsive to only ABA, the mutant promotor (with a full, active ERSE) was responsive to ABA and ER stress signalling. From these results, the authors proposed the following mechanism for interrelating ER stress with drought stress. When drought stress occurs, it activates ABA signalling, which includes a feedback mechanism for tightly regulating levels of PP2C-A and thereby confers drought tolerance. However, drought stress also activates ER stress signalling, and when PP2C-A is also upregulated by ER stress, the feedback loop is broken, causing hypersensitivity to drought. Further research is needed to see if this is indeed the mechanism by which ER components are interrelated to drought stress, though multiple lines of evidence have implicated ABA signalling in this mechanism.

## 6. ER stress signalling in response to salt stress

In addition to serving as a site for protein synthesis and transport, ER also plays a crucial role in protein homeostasis under environmental stresses [77]. Stress-induced cellular disruptions such as improper protein folding, excessive protein degradation, accumulation of unfolded protein or overexpression of a protein or increase in the signalling molecules induce ER stress [47].

Three interactive mechanisms, namely, ER quality control (ERQC), ER-associated degradation (ERAD) and the unfolded protein response (UPR) or ER-nucleus signalling pathway, have been described in relation to ER proteostasis function. Salt stress results in the accumulation of misfolded proteins leading to increased transcription of ER-localised proteins, including chaperons such as binding protein (BiP) [78] and Calreticulin (CRT) which aid in restoring proteostasis [79]. On accumulation of misfolded proteins, BiP physically binds to the misfolded proteins, prevents their aggregation and activates UPR signalling [80]. Overexpression of BiP has been reported to enhance salt stress tolerance. Wang et al. [78] identified *BiP* in Capsicum annuum L. and observed increased salt tolerance on overexpression of *CaBiP1* in *Arabidopsis*. Herath et al. [81] identified three BiP in Solanum tuberosum and observed that one of the identified candidates, StBiP3, is induced by salt stress. CRT is a $Ca^{2+}$ binding protein having molecular chaperon activity. Salt stress induces the transcript levels of CRT protein. Overexpression of CRT isoforms has been reported to increase stress tolerance in wheat [79] and tobacco [82].

ER stress signalling is also known to induce transcription of salt-responsive genes. Under salt stress, ABA levels increases which initiate proteolytic cleavage of an ER membrane-associated transcription factor bZIP17 by Golgi apparatus-resident site-1 protease (SIP). Processed bZIP17 is translocated to the nucleus to activate salt-responsive genes [83]. Liu et al. [44] showed that mutants of bZIP17 have increased sensitivity to salt stress. Ramakrishna et al. [84] reported increased salt tolerance in tobacco plant overexpressing bZIP17 from finger millet. Similarly, overexpression of ER-small heat shock protein (ER-sHSP), another ER-localised chaperon, has been reported to enhance salt tolerance in tomato (Fu et al., 2016). Guan et al. [85] characterised an ER-localised chaperon, namely Sensitive to Salt1 (SES1), which is induced by salt stress and activated by ER stress sensor bZIP17 by direct binding on the promoter region. These studies highlight the role of ER chaperones in stress signalling and acting as an important positive regulator of salt tolerance in plants.

In addition to chaperons, some enzymes involved in protein modifications also play an essential role in ERQC under salt stress. Blanco-Herrera et al. [86] observed that the mutants of UDP-glucose: glycoprotein Glucosyltransferase (UGGT), an enzyme involved in N-glycan protein modification of the misfolded proteins, are more sensitive to salt stress and negatively impacts plant growth.

Various ERAD components have been reported to act as both positive and negative regulators of salt tolerance in plants. In *Arabidopsis*, a mutant of ERAD components SEL1L/HRD3A and OS9 showed increased sensitivity to salt stress [87, 88], whereas Cui et al. [89] described a mutant for another ERAD component, ubiquitin-conjugating enzyme 32 (UBC32) to have enhanced salt tolerance in *Arabidopsis*.

Upregulation of UPR pathway genes has been associated with stress tolerance. Babitha et al. [90] reported that overexpression of a stress-responsive finger millet bZIP60 in tobacco leads to upregulation of UPR pathway genes such as BiP1, CRT1 and PDIL. The study proposed that the improved tolerance, in this case, could be through the UPR pathway.

As part of the salt stress signalling cascade, the components associated with the ERQC, ERAD and UPR mechanisms may work independently or may have regulatory connections with each other [45, 91–94]. Although many studies have reported the role of ER stress signalling in salt stress response, the underlying molecular mechanism and client proteins for the ER-associated regulatory mechanisms are still unknown. Zhang et al. [95] performed the first ER proteome analysis of wheat seedlings to understand the role of ER proteostasis pathways in salt-induced growth reduction. They proposed a putative mechanism whereby salt stress generates ROS, which triggers Redox reactions in the cell leading to the accumulation of misfolded proteins. This increase in misfolded proteins in the ER lumen then triggers ER stress which further activates UPR to relieve ER stress and maintain proteostasis.

## 7. ER stress signalling in response to cold stress

Early perception of temperature fluctuation and remodelling of the cell membrane (or lipid bilayer) is the key to acclimation to sub-optimal growth conditions. Because ER is the site of fatty acid synthesis, changes in phospholipid composition in response to cold stress are expected to be determined in the ER. In this context, Tasseva et al. [96] investigated the ability of ER membranes to alter lipid composition in *Brassica napus* and reported desaturases involved in changing membrane fluidity. Cold stress leads to unstable membrane curvatures resulting from the accumulation of diacylglycerols (DAGs) [97]. Rupiz-Lopez et al. [98] revealed that two ER-localised synaptotagmin proteins, SYT1 and SYT3, remove DAGs to prevent PM damage caused by cold stress. Additionally, ER also hosts membrane proteins that act as a sensor of cold and help elicit a stress response. In this context, Ma et al. [99] identified CHILLING TOLERANCE DIVERGENCE 1 (COLD1), a plasma membrane (PM) and ER-localised protein in rice which triggers $Ca^{2+}$ signalling in response to cold stress. Orthologs of COLD1 have also been identified in wheat [100] and Maize [101]. Overexpression of calcium-dependent protein kinases and calreticulin, also hosted by ER, has been reported to increase cold tolerance in rice [102]. William et al. [103] characterised a member of the Bcl-2-associated athanogene (BAG) family, AtBAG7 affecting cold tolerance in *Arabidopsis* is localised to the ER. Recently, Hu et al. [104] identified an ER-localised Sugars Will Eventually be Exported Transporters (SWEETs) which confers cold tolerance by regulating sugar metabolism and compartmentalisation under cold stress in *Arabidopsis*. Although many transporters situated in ER are involved in cold tolerance, our understanding of the underlying molecular mechanism is still limited.

## 8. ER stress signalling in response to biotic factors

Interactions of plants with various organisms play an essential role in its adaptability to the changing environment, not just in protecting it against various pathogens but also in enhancing its defence mechanisms [105–107]. When plants are subjected to microbial attack, pathogenesis-related (PR) proteins are synthesised as a response to it. A form of the immune response, Systemic Acquired Resistance (SAR), is established as a result of the NONEXPRESSOR OF PR GENE1 (NPR1)-regulated expression of PR genes in *Arabidopsis* [108]. Thus, it is conceivable that part of the plant immune response, including SAR, requires the synthesis and subsequent secretion of some PR proteins toward microbial pathogens. Consistent with this notion, the expression of several genes involved in protein secretion is induced in SAR of *Arabidopsis* in an NPR1- dependent manner, implying increased capability for SAR via transcriptional regulation that requires NPR1 function [109]. It was also reported in tobacco (*N. tabacum*) plants that induction of the lumenal binding protein, an ER-resident chaperone, occurs rapidly during pathogen attack and precedes the expression of PR genes [110].

Several studies have demonstrated the involvement of ER bodies in plant defence mechanisms. Even if there is an artificial induction of a wound on a plant using the wound hormone jasmonic acid, which mimics the pest chewing damage, ER bodies are stimulated [111, 112]. In a study involving Brassicaceae plants, which are resistant to arbuscular mycorrhizal fungi (AMF) symbiosis, the plants were found to be susceptible to non-AMF groups (*Piriformospora indica* in particular), which in turn enhanced the growth of the plant [113] and showed that ER bodies act in the defence mechanism [114]. *Arabidopsis* mutants were impaired in the expression of PYK10, a

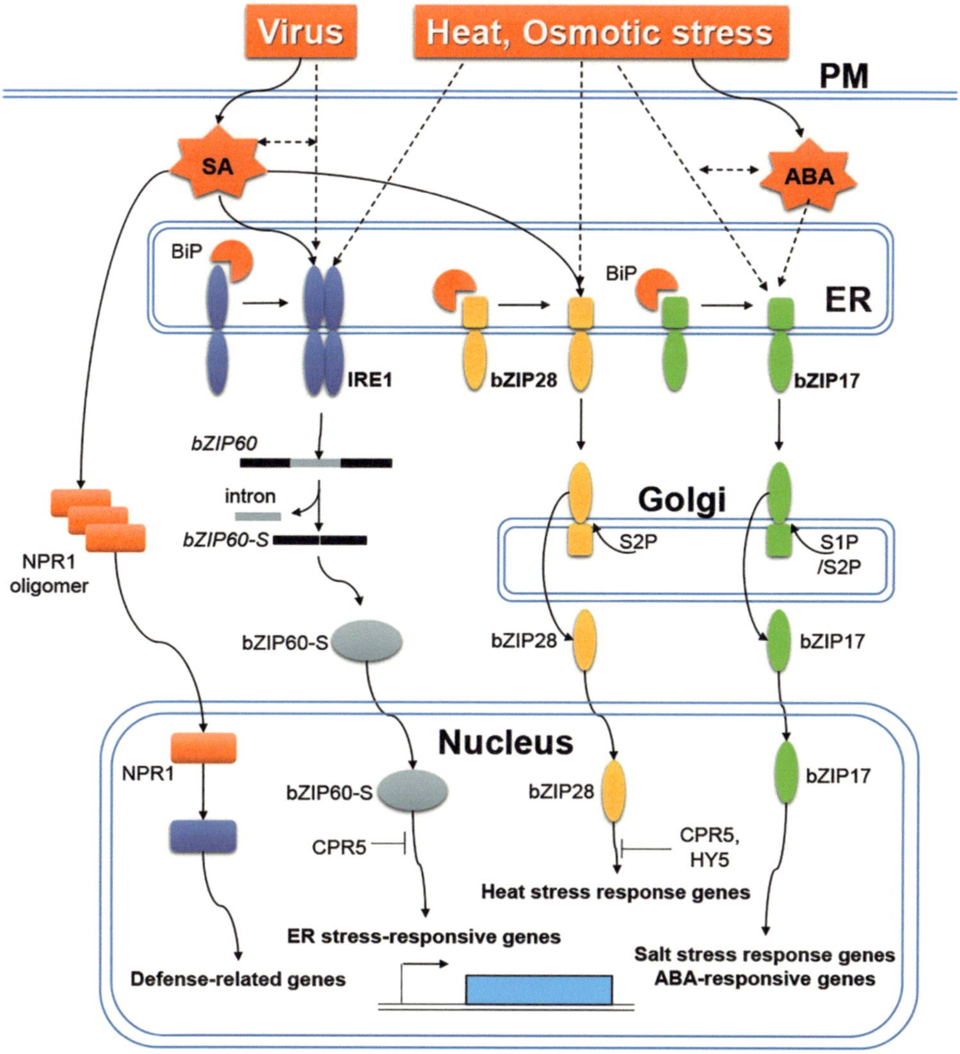


**Figure 1.**
*Summary of ER responses to biotic/abiotic stress conditions. Under biotic/abiotic stress conditions in plants such as virus infection, stress-related hormones and heat and osmotic stress, several factors including BiP, IRE1, bZIP60, bZIP17/28, genes involved in NPR1-depenedent pathways. Uncharacterized molecular pathways are indicated in dashed arrows. (adapted from [47]).*

target gene of *P. indica* in the roots of *Arabidopsis* [115]. PYK10 is a gene for an abundant myrosinase located in the ER [116, 117] which has spindle-shaped structures and is named ER bodies [27, 118, 119]. ER bodies can also be induced in rosette leaves by JA [120], and the jasmonate-insensitive coronatine insensitive1 [121] mutant does not form ER bodies [111].

## 9. Conclusions

Plants experience several stresses during their lifetime and in order to survive them, they should have a fast and dynamic response mechanism. Since nearly one third of the plant cell proteins responsible for being affected by stress are located in the ER, it has been studied that ER expresses differential responses in the plant in response to the situation. These dynamic responses in plants include changing the proteome by changing the gene expression patterns. It is the prime organelle that regulates the stress responses as any stress which affects protein folding leads to a cellular homeostatic response mediated by the UPR in response to ER stress. **Figure 1** highlights critical ER responses against perceived biotic/abiotic stress conditions.

## Acknowledgements

V.J.B. and B.S. thankfully acknowledge the Department of Biotechnology, Govt. of India for providing research grant to Assam Agricultural University and Dibrugarh University [Grant No. BT/PR25099/NER/95/1014/2017, dated 29 May 2018 and Grant no. BT/PR36255/NNT/28/1728/2020, dated 01/12/2021]. V.J.B. also acknowledge the support rendered by DBT e-Library Consortium (DeLCON) to Centre for Biotechnology and Bioinformatics at Dibrugarh University.

## Conflict of interest

The authors declare no conflict of interest.

*DOI: http://dx.doi.org/10.5772/intechopen.106590*

**Author details**

Vishwa Jyoti Baruah[1]*, Bhaswati Sarmah[2,3], Manny Saluja[4,5] and Elizabeth H. Mahood[3]

1 Centre for Biotechnology and Bioinformatics, Faculty of Biological Sciences, Dibrugarh University, Dibrugarh, Assam, India

2 Department of Plant Breeding and Genetics, Assam Agricultural University, Jorhat, Assam, India

3 Plant Biology Section, School of Integrative Plant Science, Cornell University, Ithaca, NY, USA

4 KWS Gateway Research Center, LLC (KWS), St. Louis, MO, USA

5 Boyce Thompson Institute, Ithaca, NY, USA

*Address all correspondence to: vishwabaruah@gmail.com

## References

[1] Schuldiner M, Schwappach B. From rags to riches - the history of the endoplasmic reticulum. Biochim Biophys Acta - Mol. Cell Research. 2013;**1833**(11):2389-2391. [Internet]. DOI: 10.1016%2Fj.bbamcr.2013.03.005

[2] Voeltz GK, Rolls MM, Rapoport TA. Structural organization of the endoplasmic reticulum. EMBO Reports. 2002;**3**(10):944-950. [Internet]. DOI: 10.1093%2Fembo-reports% 2Fkvf202

[3] Westrate LM, Lee JE, Prinz WA, Voeltz GK. Form follows function: The importance of endoplasmic reticulum shape. Annual Review of Biochemistry. 2015;**84**(1):791-811. [Internet]. DOI: 10.1146%2Fannurev-biochem-072711-163501

[4] Staehelin LA. The plant ER: A dynamic organelle composed of a large number of discrete functional domains. The Plant Journal. 1997;**11**(6):1151-1165. [Internet]. DOI: 10.1046%2Fj.1365-313x.1997.11061151.x

[5] Aridor M, Hannan LA. Tarffic jam: A compendium of human diseases that affect intracellular transport processes. Traffic 2000;1(11):836-851. [Internet]: DOI:10.1034%2Fj.1600-0854.2000.011104.x

[6] Kim SJ, Brandizzi F. The plant secretory pathway for the trafficking of cell wall polysaccharides and glycoproteins. Glycobiology. 2016;**26**(9):940-949. [Internet]. DOI: 10.1093%2Fglycob%2Fcww044

[7] Brandizzi F. Transport from the endoplasmic reticulum to the Golgi in plants: Where are we now? Seminars in Cell & Developmental Biology. 2018;**80**:94-105. [Internet]. DOI: 10.1016%2Fj.semcdb.2017.06.024

[8] Vitale A, Denecke J. The endoplasmic reticulum - gateway of the secretory pathway. The Plant Cell. 1999;**11**(4):615-628. [Internet]. DOI: 10.1105% 2Ftpc.11.4.615

[9] Vitale A, Galili G. The endomembrane system and the problem of protein sorting. Plant Physiology. 2001;**125**(1):115-118. [Internet]. DOI: 10.1104%2Fpp.125.1.115

[10] Angelos E, Ruberti C, Kim SJ, Brandizzi F. Maintaining the factory: The roles of the unfolded protein response in cellular homeostasis in plants. The Plant Journal. 2017;**90**(4):671-682. [Internet]. DOI: 10.1111%2Ftpj.13449

[11] Brandizzi F, Snapp EL, Roberts AG, Lippincott-Schwartz J, Hawes C. Membrane protein transport between the endoplasmic reticulum and the Golgi in tobacco leaves is energy dependent but cytoskeleton independent: Evidence from selective photobleaching. The Plant Cell. 2002;**14**(6):1293-1309. [Internet]. DOI: 10.1105%2Ftpc. 001586

[12] Shimada T, Fuji K, Tamura K, Kondo M, Nishimura M, Hara-Nishimura I. Vacuolar sorting receptor for seed storage proteins in Arabidopsis thaliana. Proceedings of the National Academy of Sciences of the United States of America. 2003;**100**(26):16095-16100. [Internet]. DOI: 10.1073%2Fpnas.2530568100

[13] Hurlock AK, Roston RL, Wang K, Benning C. Lipid trafficking in plant cells. Traffic. 2014;**15**(9):915-932. [Internet]. DOI: 10.1111%2Ftra.12187

[14] Block MA, Jouhet J. Lipid trafficking at endoplasmic reticulum-chloroplast membrane contact sites. Current Opinion in Cell Biology. 2015;**35**:21-29. [Internet]. DOI: 10.1016%2Fj.ceb.2015.03.004

[15] Tamura K, Shimada T, Kondo M, Nishimura M, Hara-Nishimura I. KATAMARI1/MURUS3 is a novel Golgi membrane protein that is required for endomembrane organization in Arabidopsis. The Plant Cell. 2005;**17**(6):1764-1776. [Internet]. DOI: 10.1105%2Ftpc.105.031930

[16] Conger R, Chen Y, Fornaciari S, Faso C, Held MA, Renna L, et al. Evidence for the involvement of the Arabidopsis SEC24A in male transmission. Journal of Experimental Botany. 2011;**62**(14):4917-4926. [Internet]. DOI: 10.1093%2Fjxb%2Ferr174

[17] Stefano G, Renna L, Moss T, McNew JA, Brandizzi F. In Arabidopsis, the spatial and dynamic organization of the endoplasmic reticulum and Golgi apparatus is influenced by the integrity of the C-terminal domain of RHD3, a non-essential GTPase. The Plant Journal. 2012;**69**(6):957-966. [Internet]. DOI: 10.1111%2Fj.1365-313x.2011.04846.x

[18] Renna L, Stefano G, Majeran W, Micalella C, Meinnel T, Giglione C, et al. Golgi traffic and integrity depend on N-myristoyl transferase-1 in Arabidopsis. The Plant Cell. 2013;**25**(5):1756-1773. [Internet]. DOI: 10.1105%2Ftpc.113.111393

[19] Hawes CR, Juniper BE, Horne JC. Low and high voltage electron microscopy of mitosis and cytokinesis in maize roots. Planta. 1981;**152**(5):397-407. [Internet]. DOI: 10.1007%2Fbf00385355

[20] Craig S, Staehelin LA. High pressure freezing of intact plant tissues. Evaluation and characterization of novel features of the endoplasmic reticulum and associated membrane systems. European Journal of Cell Biology. 1988;**46**(1):81-93. [Internet] Available from:: http://www.ncbi.nlm.nih.gov/pubmed/3396590

[21] Wright KM, Oparka KJ. The ER within plasmodesmata. In: Plant Cell Monographs. Berlin Heidelberg: Springer; 2006. pp. 279-308. [Internet]. DOI: 10.1007%2F7089_060

[22] Carr DJ. Historical perspectives on Plasmodesmata. In: Intercellular Communication in Plants: Studies on Plasmodesmata. Berlin Heidelberg: Springer; 1976. pp. 291-295. [Internet]. DOI: 10.1007%2F978-3-642-66294-2_14

[23] Juniper BE, Hawes CR, Horne JC. The relationships between the Dictyosomes and the forms of endoplasmic reticulum in plant cells with different export programs. Botanical Gazette. 1982;**143**(2):135-145. [Internet]. DOI: 10.1086%2F337282

[24] Andersson MX, Goksör M, Sandelius AS. Optical manipulation reveals strong attracting forces at membrane contact sites between endoplasmic reticulum and chloroplasts. The Journal of Biological Chemistry. 2007;**282**(2):1170-1174. [Internet]. DOI: 10.1074%2Fjbc.m608124200

[25] Sparkes I, Hawes C, Frigerio L. FrontiERs: Movers and shapers of the higher plant cortical endoplasmic reticulum. Current Opinion in Plant Biology. 2011;**14**(6):658-665. [Internet]. DOI: 10.1016%2Fj.pbi.2011.07.006

[26] Lai YS, Stefano G, Brandizzi F. ER stress signaling requires RHD3, a

functionally conserved ER-shaping GTPase. Journal of Cell Science. 2014;**127**(15):3227-3232. [Internet]. DOI: 10.1242%2Fjcs.147447

[27] Ridge RW, Uozumi Y, Plazinski J, Hurley UA, Williamson RE. Developmental transitions and dynamics of the cortical ER of Arabidopsis cells seen with green fluorescent protein. Plant & Cell Physiology. 1999;**40**(12):1253-1261. [Internet]. DOI: 10.1093%2Foxfordjournals.pcp.a029513

[28] Sparkes I, Runions J, Hawes C, Griffing L. Movement and remodeling of the endoplasmic reticulum in nondividing cells of tobacco leaves. The Plant Cell. 2009;**21**(12):3937-3949. [Internet]. DOI: 10.1105%2Ftpc.109.072249

[29] Stefano G, Renna L, Brandizzi F. The endoplasmic reticulum exerts control over organelle streaming during cell expansion. Journal of Cell Science. 2014;**127**(5):947-953. [Internet]. DOI: 10.1242%2Fjcs.139907

[30] Stefano G, Renna L, Lai Y, Slabaugh E, Mannino N, Buono RA, et al. ER network homeostasis is critical for plant endosome streaming and endocytosis. Cell Discovery. 2015; Nov;**1**(1):1-16. [Internet]. DOI: 10.1038%2Fcelldisc.2015.33

[31] McFarlane HE, Lee EK, Van Bezouwen LS, Ross B, Rosado A, Samuels AL. Multiscale structural analysis of plant ER-PM contact sites. Plant & Cell Physiology. 2017;**58**(3):478-484. [Internet]. DOI: 10.1093%2Fpcp%2Fpcw224

[32] Hepler PK, Palevitz BA, Lancelle SA, McCauley MM, Lichtscheidl I. Cortical endoplasmic reticulum in plants. Journal of Cell Science. 1990;**96**(3):355-373. [Internet]. DOI: 10.1242%2Fjcs.96.3.355

[33] Quader H, Zachariadis M. The morphology and dynamics of the ER. In: Plant Cell Monographs. Berlin Heidelberg: Springer; 2006. pp. 1-23. [Internet]. DOI: 10.1007%2F7089_063

[34] Pandey P, Ramegowda V, Senthil-Kumar M. Shared and unique responses of plants to multiple individual stresses and stress combinations: Physiological and molecular mechanisms. Frontiers in Plant Science. [Internet]. 2015 Sep;6:723. https://doi.org/10.3389%2Ffpls.2015.00723

[35] Zhu JK. Abiotic stress signaling and responses in plants. Cell. 2016;**167**(2):313-324. [Internet]. DOI: 10.1016%2Fj.cell.2016.08.029

[36] Ellgaard L, Helenius A. Quality control in the endoplasmic reticulum. Nature Reviews. Molecular Cell Biology. 2003;**4**(3):181-191. [Internet]. DOI: 10.1038%2Fnrm1052

[37] Schröder M, Kaufman RJ. The mammalian unfolded protein response. Annual Review of Biochemistry. 2005;**74**(1):739-789. [Internet]. DOI: 10.1146%2Fannurev.biochem.73.011303.074134

[38] Fontes EBP, Shank BB, Wrobel RL, Moose SP, OBrian GR, Wurtzel ET, et al. Characterization of an immunoglobulin binding protein homolog in the maize floury-2 endosperm mutant. The Plant Cell. 1991;**3**(5):483-496. [Internet]. DOI: 10.1105%2Ftpc.3.5.483

[39] Strasser R. Protein quality control in the endoplasmic reticulum of plants. Annual Review of Plant Biology. 2018;**69**(1):147-172. [Internet]. DOI: 10.1146%2Fannurev-arplant-042817-040331

[40] Chen Y, Zhang Y, Yin Y, Gao G, Li S, Jiang Y, et al. SPD - a web-based secreted

protein database. Nucleic Acids Research. 2005;**33**(DATABASE ISS):D169-D173. [Internet]. DOI: 10.1093%2Fnar%2Fgki093

[41] Howell SH. Endoplasmic reticulum stress responses in plants. Annual Review of Plant Biology. 2013;**64**(1):477-499. [Internet]. DOI: 10.1146%2Fannurev-arplant-050312-120053

[42] Al-Whaibi MH. Plant heat-shock proteins: A mini review. Journal of King Saudi Arabia University - Science. 2011;**23**(2):139-150. [Internet]. DOI: 10.1016%2Fj.jksus.2010.06.022

[43] Gao H, Brandizzi F, Benning C, Larkin RM. A membrane-tethered transcription factor defines a branch of the heat stress response in Arabidopsis thaliana. Proceedings of the National Academy of Sciences of the United States of America. 2008;**105**(42):16397-16403. [Internet]. DOI: 10.1073%2Fpnas.0808463105

[44] Liu JX, Srivastava R, Che P, Howell SH. Salt stress responses in Arabidopsis utilize a signal transduction pathway related to endoplasmic reticulum stress signaling. The Plant Journal. 2007;**51**(5):897-909. [Internet]. DOI: 10.1111%2Fj.1365-313x.2007.03195.x

[45] Che P, Bussell JD, Zhou W, Estavillo GM, Pogson BJ, Smith SM. Signaling from the endoplasmic reticulum activates brassinosteroid signaling and promotes acclimation to stress in Arabidopsis. Science Signaling. 2010;**3**(141):ra69-ra69. [Internet]. DOI: 10.1126%2Fscisignal.2001140

[46] Deng Y, Humbert S, Liu JX, Srivastava R, Rothstein SJ, Howell SH. Heat induces the splicing by IRE1 of a mRNA encoding a transcription factor involved in the unfolded protein response in Arabidopsis. Proceedings of the National Academy of Sciences of the United States of America. 2011;**108**(17):7247-7252. [Internet]. DOI: 10.1073%2Fpnas.1102117108

[47] Park CJ, Park JM. Endoplasmic reticulum plays a critical role in integrating signals generated by both biotic and abiotic stress in plants. Frontiers in Plant Science. 2019;**10**:399. [Internet]. DOI: 10.3389%2Ffpls.2019.00399

[48] Wakasa Y, Yasuda H, Oono Y, Kawakatsu T, Hirose S, Takahashi H, et al. Expression of er quality control-related genes in response to changes in BiP1 levels in developing rice endosperm. The Plant Journal. 2011;**65**(5):675-689. [Internet]. DOI: 10.1111%2Fj.1365-313x.2010.04453.x

[49] Yasuda H, Hirose S, Kawakatsu T, Wakasa Y, Takaiwa F. Overexpression of BiP has inhibitory effects on the accumulation of seed storage proteins in endosperm cells of rice. Plant & Cell Physiology. 2009;**50**(8):1532-1543. [Internet]. DOI: 10.1093%2Fpcp%2Fpcp098

[50] Sandhu J, Irvin L, Liu K, Staswick P, Zhang C, Walia H. Endoplasmic reticulum stress pathway mediates the early heat stress response of developing rice seeds. Plant, Cell & Environment. 2021;**44**(8):2604-2624. [Internet]. DOI: 10.1111%2Fpce.14103

[51] Ishimaru T, Parween S, Saito Y, Shigemitsu T, Yamakawa H, Nakazono M, et al. Laser microdissection-based tissue-specific transcriptome analysis reveals a novel regulatory network of genes involved in heat-induced grain chalk in Rice endosperm. Plant & Cell Physiology. 2019;**60**(3):626-642. [Internet]. DOI: 10.1093%2Fpcp%2Fpcy233

[52] Mueller SP, Unger M, Guender L, Fekete A, Mueller MJ. Phospholipid: Diacylglycerol acyltransferase-mediated triacylglyerol synthesis augments basal thermotolerance. Plant Physiology. 2017;175(1):486-497. [Internet]: DOI: 10.1104%2Fpp.17.00861

[53] Spicher L, Glauser G, Kessler F. Lipid antioxidant and galactolipid remodeling under temperature stress in tomato plants. Frontiers in Plant Science. 2016;**7**:167. (FEB2016) [Internet]. DOI: 10.3389%2Ffpls.2016.00167

[54] Narayanan S, Tamura PJ, Roth MR, Prasad PVV, Welti R. Wheat leaf lipids during heat stress: I. high day and night temperatures result in major lipid alterations. Plant, Cell & Environment. 2016;**39**(4):787-803. [Internet]. DOI: 10.1111%2Fpce.12649

[55] Zhang X, Xu Y, Huang B. Lipidomic reprogramming associated with drought stress priming-enhanced heat tolerance in tall fescue (Festuca arundinacea). Plant, Cell & Environment. 2019;**42**(3):947-958. [Internet]. DOI: 10.1111%2Fpce.13405

[56] Martinière A, Shvedunova M, Thomson AJW, Evans NH, Penfield S, Runions J, et al. Homeostasis of plasma membrane viscosity in fluctuating temperatures. The New Phytologist. 2011;**192**(2):328-337. [Internet]. DOI: 10.1111%2Fj.1469-8137.2011.03821.x

[57] Nguyen VC, Nakamura Y, Kanehara K. Membrane lipid polyunsaturation mediated by FATTY ACID DESATURASE 2 (FAD2) is involved in endoplasmic reticulum stress tolerance in Arabidopsis thaliana. The Plant Journal. 2019;**99**(3):478-493. [Internet]. DOI: 10.1111%2Ftpj.14338

[58] Tang GQ, Novitzky WP, Carol Griffin H, Huber SC, Dewey RE. Oleate desaturase enzymes of soybean: Evidence of regulation through differential stability and phosphorylation. The Plant Journal. 2005;**44**(3):433-446. [Internet]. DOI: 10.1111%2Fj.1365-313x.2005.02535.x

[59] Higashi Y, Okazaki Y, Myouga F, Shinozaki K, Saito K. Landscape of the lipidome and transcriptome under heat stress in Arabidopsis thaliana. Scientific Reports. 2015;**5**(1):1-11 [Internet]. DOI: 10.1038%2Fsrep10533

[60] Saidi Y, Finka A, Muriset M, Bromberg Z, Weiss YG, Maathuis FJM, et al. The heat shock response in moss plants is regulated by specific calcium-permeable channels in the plasma membrane. Plant Cell. 2009;**21**(9):2829-2843. [Internet]. DOI: 10.1105%2Ftpc.108.065318

[61] Bokszczanin KL, Fragkostefanakis S. Perspectives on deciphering mechanisms underlying plant heat stress response and thermotolerance. Frontiers in Plant Science. 2013;**4**:315. (AUG) [Internet]. DOI: 10.3389%2Ffpls.2013.00315

[62] Nawkar GM, Kang CH, Maibam P, Park JH, Jung YJ, Chae HB, et al. HY5, a positive regulator of light signaling, negatively controls the unfolded protein response in Arabidopsis. Proceedings of the National Academy of Sciences of the United States of America. 2017;**114**(8):2084-2089. [Internet]. DOI: 10.1073%2Fpnas.1609844114

[63] Jung JH, Domijan M, Klose C, Biswas S, Ezer D, Gao M, et al. Phytochromes function as thermosensors in Arabidopsis. Science (80- ). 2016;**354**(6314):886-889. [Internet]. DOI: 10.1126%2Fscience.aaf6005

[64] Deng Y, Srivastava R, Quilichini TD, Dong H, Bao Y, Horner HT, et al. IRE1, a component of the unfolded protein

response signaling pathway, protects pollen development in Arabidopsis from heat stress. The Plant Journal. 2016;**88**(2):193-204. [Internet]. DOI: 10.1111%2Ftpj.13239

[65] Rutley N, Poidevin L, Doniger T, Tillett RL, Rath A, Forment J, et al. Characterization of novel pollen-expressed transcripts reveals their potential roles in pollen heat stress response in Arabidopsis thaliana. Plant Reproduction. 2021;**34**(1):61-78. [Internet]. DOI: 10.1007%2Fs00497-020-00400-1

[66] Yamamoto M, Uji S, Sugiyama T, Sakamoto T, Kimura S, Endo T, et al. ERdj3B-mediated quality control maintains anther development at high temperatures. Plant Physiology. 2020;**182**(4):1979-1990. [Internet]. DOI: 10.1104%2Fpp.19.01356

[67] Lambers H, Chapin FS, Pons TL. Photosynthesis. In: Plant Physiological Ecology. New York: Springer; 2008. pp. 11-99. [Internet]. DOI: 10.1007%2F978-0-387-78341-3_2

[68] Yamaguchi-Shinozaki K, Shinozaki K. Transcriptional regulatory networks in cellular responses and tolerance to dehydration and cold stresses. Annual Review of Plant Biology. 2006;**57**(1):781-803. [Internet]. DOI: 10.1146%2Fannurev.arplant.57.032905.105444

[69] Valente MAS, Faria JAQA, Soares-Ramos JRL, Reis PAB, Pinheiro GL, Piovesan ND, et al. The ER luminal binding protein (BiP) mediates an increase in drought tolerance in soybean and delays drought-induced leaf senescence in soybean and tobacco. Journal of Experimental Botany. 2009;**60**(2):533-546. [Internet]. DOI: 10.1093%2Fjxb%2Fern296

[70] Alvim FC, Carolino SMB, Cascardo JCM, Nunes CC, Martinez CA, Otoni WC, et al. Enhanced accumulation of BiP in transgenic plants confers tolerance to water stress. Plant Physiology. 2001;**126**(3):1042-1054. [Internet]. DOI: 10.1104%2Fpp.126.3.1042

[71] Foresti O, Ruggiano A, Hannibal-Bach HK, Ejsing CS, Carvalho P. Sterol homeostasis requires regulated degradation of squalene monooxygenase by the ubiquitin ligase Doa10/Teb4. eLife. 2013;**2013**(2):e00953 [Internet]. DOI: 10.7554%2Felife.00953

[72] Doblas VG, Amorim-Silva V, Posé D, Rosado A, Esteban A, Arró M, et al. The SUD1 gene encodes a putative E3 ubiquitin ligase and is a positive regulator of 3-hydroxy-3-methylglutaryl coenzyme a reductase activity in Arabidopsis. Plant Cell. 2013;**25**(2):728-743. [Internet]. DOI: 10.1105%2Ftpc.112.108696

[73] Posé D, Castanedo I, Borsani O, Nieto B, Rosado A, Taconnat L, et al. Identification of the Arabidopsis dry2/sqe1-5 mutant reveals a central role for sterols in drought tolerance and regulation of reactive oxygen species. The Plant Journal. 2009;**59**(1):63-76. [Internet]. DOI: 10.1111%2Fj.1365-313x.2009.03849.x

[74] Lee HK, Cho SK, Son O, Xu Z, Hwang I, Kim WT. Drought stress-induced Rma1H1, a RING membrane-anchor E3 ubiquitin ligase homolog, regulates aquaporin levels via ubiquitination in transgenic arabidopsis plants. The Plant Cell. 2009;**21**(2):622-641. [Internet]. DOI: 10.1105%2Ftpc.108.061994

[75] Wang B, Du H, Zhang Z, Xu W, Deng X. BhbZIP60 from resurrection plant Boea hygrometrica is an mRNA splicing-activated endoplasmic reticulum

stress regulator involved in drought tolerance. Frontiers in Plant Science 2017;**8**:245. [Internet]. DOI: 10.3389% 2Ffpls.2017.00245

[76] Xiang Y, Sun X, Gao S, Qin F, Dai M. Deletion of an endoplasmic reticulum stress response element in a ZmPP2C-A gene facilitates drought tolerance of maize seedlings. Molecular Plant. 2017;**10**(3):456-469. [Internet]. DOI: 10.1016%2Fj.molp.2016.10.003

[77] Reyes-Impellizzeri S, Moreno AA. The endoplasmic reticulum role in the plant response to abiotic stress. Frontiers in Plant Science. 2021;**12**:2582. [Internet]. DOI: 10.3389%2Ffpls.2021.755447

[78] Wang H, Niu H, Zhai Y, Lu M. Characterization of BiP genes from pepper (Capsicum annuum L.) and the role of CaBiP1 in response to endoplasmic reticulum and multiple abiotic stresses. Frontiers in Plant Science. 2017;**8**:1122. [Internet]. DOI: 10.3389% 2Ffpls.2017.01122

[79] Wang J, Li R, Mao X, Jing R. Functional analysis and marker development of TaCRT-D gene in common wheat (Triticum aestivum L.). Frontiers in Plant Science. 2017;**8**:1557. [Internet]. DOI: 10.3389%2Ffpls. 2017.01557

[80] Srivastava R, Deng Y, Shah S, Rao AG, Howell SH. Binding protein is a master regulator of the endoplasmic reticulum stress sensor/transducer bZIP28 in Arabidopsis. The Plant Cell. 2013;**25**(4):1416-1429. [Internet]. DOI: 10.1105%2Ftpc.113.110684

[81] Herath V, Gayral M, Adhikari N, Miller R, Verchot J. Genome-wide identification and characterization of Solanum tuberosum BiP genes reveal the role of the promoter architecture in BiP gene diversity. Scientific Reports. 2020;**10**(1):1-4 [Internet]. DOI: 10.1038%2Fs41598-020-68407-2

[82] Xiang Y, Hai LY, Song M, Wang Y, Xu W, Wu L, et al. Overexpression of a triticum aestivum calreticulin gene (TaCRT1) improves salinity tolerance in tobacco. Zhang JS, editor. PLoS One. 2015;**10**(10):e0140591. [Internet]. DOI: 10.1371%2Fjournal.pone.0140591

[83] Zhou SF, Sun L, Valdés AE, Engström P, Song ZT, Lu SJ, et al. Membrane-associated transcription factor peptidase, site-2 protease, antagonizes ABA signaling in Arabidopsis. The New Phytologist. 2015;**208**(1):188-197. [Internet]. DOI: 10.1111%2Fnph.13436

[84] Ramakrishna C, Singh S, Raghavendrarao S, Padaria JC, Mohanty S, Sharma TR, et al. The membrane tethered transcription factor EcbZIP17 from finger millet promotes plant growth and enhances tolerance to abiotic stresses. Scientific Reports. 2018;**8**(1):1-4 [Internet]. DOI: 10.1038%2Fs41598-018-19766-4

[85] Guan P, Wang J, Li H, Xie C, Zhang S, Wu C, et al. SENSITIVE TO SALT1, an endoplasmic reticulum-localized chaperone, positively regulates SALT resistance. Plant Physiology. 2018;**178**(3):1390-1405. [Internet]. DOI: 10.1104%2Fpp.18.00840

[86] Blanco-Herrera F, Moreno AA, Tapia R, Reyes F, Araya M, D'Alessio C, et al. The UDP-glucose: Glycoprotein glucosyltransferase (UGGT), a key enzyme in ER quality control, plays a significant role in plant growth as well as biotic and abiotic stress in Arabidopsis thaliana. BMC Plant Biology. 2015;**15**(1):1-2 [Internet]. DOI: 10.1186%2Fs12870-015-0525-2

[87] Hüttner S, Veit C, Schoberer J, Grass J, Strasser R. Unraveling the function of Arabidopsis thaliana OS9 in the endoplasmic reticulum-associated degradation of glycoproteins. Plant Molecular Biology. 2012;**79**(1-2):21-33. [Internet]. DOI: 10.1007%2Fs11103-012-9891-4

[88] Liu L, Cui F, Li Q, Yin B, Zhang H, Lin B, et al. The endoplasmic reticulum-associated degradation is necessary for plant salt tolerance. Cell Research. 2011;**21**(6):957-969. [Internet]. DOI: 10.1038%2Fcr.2010.181

[89] Cui F, Liu L, Zhao Q, Zhang Z, Li Q, Lin B, et al. Arabidopsis ubiquitin conjugase UBC32 is an ERAD component that functions in brassinosteroid-mediated salt stress tolerance. The Plant Cell. 2012;**24**(1):233-244. [Internet]. DOI: 10.1105%2Ftpc.111.093062

[90] Babitha KC, Ramu SV, Nataraja KN, Sheshshayee MS, Udayakumar M. EcbZIP60, a basic leucine zipper transcription factor from Eleusine coracana L. improves abiotic stress tolerance in tobacco by activating unfolded protein response pathway. Molecular Breeding. 2015;**35**(9):1-7 [Internet]. DOI: 10.1007%2Fs11032-015-0374-6

[91] Fujita M, Mizukado S, Fujita Y, Ichikawa T, Nakazawa M, Seki M, et al. Identification of stress-tolerance-related transcription-factor genes via mini-scale full-length cDNA over-eXpressor (FOX) gene hunting system. Biochemical and Biophysical Research Communications. 2007;**364**(2):250-257. [Internet]. DOI: 10.1016%2Fj.bbrc.2007.09.124

[92] Henriquez-Valencia C, Moreno AA, Sandoval-Ibañez O, Mitina I, Blanco-Herrera F, Cifuentes-Esquivel N, et al. BZIP17 and bZIP60 regulate the expression of BiP3 and other salt stress responsive genes in an UPR-independent manner in Arabidopsis thaliana. Journal of Cellular Biochemistry. 2015;**116**(8):1638-1645. [Internet]. DOI: 10.1002%2Fjcb.25121

[93] Tian L, Zhang Y, Kang E, Ma H, Zhao H, Yuan M, et al. Basic-leucine zipper 17 and Hmg-CoA reductase degradation 3A are involved in salt acclimation memory in Arabidopsis. Journal of Integrative Plant Biology. 2019;**61**(10):1062-1084. [Internet]. DOI: 10.1111%2Fjipb.12744

[94] Li Q, Wei H, Liu L, Yang X, Zhang X, Xie Q. Unfolded protein response activation compensates endoplasmic reticulum-associated degradation deficiency in Arabidopsis. Journal of Integrative Plant Biology. 2017;**59**(7):506-521. [Internet]. DOI: 10.1111%2Fjipb.12544

[95] Zhang J, Liu D, Zhu D, Liu N, Yan Y. Endoplasmic reticulum subproteome analysis reveals underlying defense mechanisms of wheat seedling leaves under salt stress. International Journal of Molecular Sciences. 2021;**22**(9):4840. [Internet]. DOI: 10.3390%2Fijms22094840

[96] Tasseva G, De Virville JD, Cantrel C, Moreau F, Zachowski A. Changes in the endoplasmic reticulum lipid properties in response to low temperature in Brassica napus. Plant Physiology and Biochemistry. 2004;**42**(10):811-822. [Internet]. DOI: 10.1016%2Fj.plaphy.2004.10.001

[97] Carrasco S, Mérida I. Diacylglycerol, when simplicity becomes complex. Trends in Biochemical Sciences. 2007;**32**(1):27-36. [Internet]. DOI: 10.1016%2Fj.tibs.2006.11.004

[98] Ruiz-Lopez N, Pérez-Sancho J, del Valle AE, Haslam RP, Vanneste S,

Catalá R, et al. Synaptotagmins at the endoplasmicreticulum–plasmamembrane contact sites maintain diacylglycerol homeostasis during abiotic stress. The Plant Cell. 2021;**33**(7):2431-2453. [Internet]. DOI: 10.1093%2Fplcell%2Fkoab122

[99] Ma Y, Dai X, Xu Y, Luo W, Zheng X, Zeng D, et al. COLD1 confers chilling tolerance in rice. Cell. 2015;**160**(6):1209-1221. [Internet]. DOI: 10.1016%2Fj.cell.2015.01.046

[100] Dong H, Yan S, Liu J, Liu P, Sun J. TaCOLD1 defines a new regulator of plant height in bread wheat. Plant Biotechnology Journal. 2019;**17**(3):687-699. [Internet]. DOI: 10.1111%2Fpbi.13008

[101] Jin YN, Cui Z h, Ma K, Yao JL, Ruan YY, Guo ZF. Characterization of ZmCOLD1, novel GPCR-type G protein genes involved in cold stress from Zea mays L. and the evolution analysis with those from other species. Physiology and Molecular Biology of Plants. 2021;**27**(3):619-632. [Internet]. DOI: 10.1007%2Fs12298-021-00966-8

[102] Komatsu S, Yang G, Khan M, Onodera H, Toki S, Yamaguchi M. Over-expression of calcium-dependent protein kinase 13 and calreticulin interacting protein 1 confers cold tolerance on rice plants. Molecular Genetics and Genomics. 2007;**277**(6):713-723. [Internet]. DOI: 10.1007%2Fs00438-007-0220-6

[103] Williams B, Kabbage M, Britt R, Dickman MB. AtBAG7, an Arabidopsis Bcl-2-associated athanogene, resides in the endoplasmic reticulum and is involved in the unfolded protein response. Proceedings of the National Academy of Sciences of the United States of America. 2010;**107**(13):6088-6093. [Internet]. DOI: 10.1073%2Fpnas.0912670107

[104] Hu L, Zhang F, Song S, Yu X, Ren Y, Zhao X, et al. CsSWEET2, a hexose transporter from cucumber (Cucumis sativus L.), affects sugar metabolism and improves cold tolerance in Arabidopsis. International Journal of Molecular Sciences. 2022;**23**(7):3886. [Internet]. DOI: 10.3390%2Fijms23073886

[105] Lau JA, Lennon JT. Rapid responses of soil microorganisms improve plant fitness in novel environments. Proceedings of the National Academy of Sciences of the United States of America. 2012;**109**(35):14058-14062. [Internet]. DOI: 10.1073%2Fpnas.1202319109

[106] Finkel OM, Castrillo G, Herrera Paredes S, Salas González I, Dangl JL. Understanding and exploiting plant beneficial microbes. Current Opinion in Plant Biology. 2017;**38**:155-163. [Internet]. DOI: 10.1016%2Fj.pbi.2017.04.018

[107] Parniske M. Arbuscular mycorrhiza: The mother of plant root endosymbioses. Nature Reviews. Microbiology. 2008;6(10):763-775. [Internet]: DOI: 10.1038%2Fnrmicro1987

[108] Durrant WE, Dong X. Systemic acquired resistance. Annual Review of Phytopathology. 2004;**42**(1):185-209. [Internet]. DOI: 10.1146%2Fannurev.phyto.42.040803.140421

[109] Wang D, Weaver ND, Kesarwani M, Dong X. Induction of protein secretory pathway is required for systemic acquired resistance. Science (80-). 2005;**308**(5724):1036-1040. [Internet]. DOI: 10.1126%2Fscience.1108791

[110] Jelitto-Van Dooren EPWM, Vidal S, Denecke J. Anticipating endoplasmic reticulum stress: A novel early response before pathogenesis-related gene induction. The Plant Cell. 1999;**11**(10):1935-1943. [Internet]. DOI: 10.1105%2Ftpc.11.10.1935

[111] Matsushima R, Hayashi Y, Kondo M, Shimada T, Nishimura M, Hara-Nishimura I. An endoplasmic reticulum-derived structure that is induced under stress conditions in Arabidopsis. Plant Physiology. 2002;**130**(4):1807-1814. [Internet]. DOI: 10.1104%2Fpp.009464

[112] Hara-Nishimura I, Matsushima R. A wound-inducible organelle derived from endoplasmic reticulum: A plant strategy against environmental stresses? Current Opinion in Plant Biology. 2003;**6**(6):583-588. [Internet]. DOI: 10.1016%2Fj.pbi.2003.09.015

[113] Kumari R, Kishan H, Bhoon YK, Varma A. Colonization of cruciferous plants by Piriformospora indica. Current Science. 2003;**85**(12):1672-1674 [cited 2022]; [Internet]. Available from: http://www.jstor.org/stable/24109969

[114] Sherameti I, Venus Y, Drzewiecki C, Tripathi S, Dan VM, Nitz I, et al. PYK10, a β-glucosidase located in the endoplasmatic reticulum, is crucial for the beneficial interaction between Arabidopsis thaliana and the endophytic fungus Piriformospora indica. The Plant Journal. 2008;**54**(3):428-439. [Internet]. DOI: 10.1111%2Fj.1365-313x.2008.03424.x

[115] Peškan-Berghöfer T, Shahollari B, Pham HG, Hehl S, Markert C, Blanke V, et al. Association of Piriformospora indica with Arabidopsis thaliana roots represents a novel system to study beneficial plant-microbe interactions and involves early plant protein modifications in the endoplasmic reticulum and at the plasma membrane. Physiologia Plantarum. 2004;**122**(4):465-477. [Internet]. DOI: 10.1111%2Fj.1399-3054.2004.00424.x

[116] Nitz I, Berkefeld H, Puzio PS, Grundler FMW. Pyk10, a seedling and root specific gene and promoter from Arabidopsis thaliana. Plant Science. 2001;**161**(2):337-346. [Internet]. DOI: 10.1016%2Fs0168-9452%2801%2900412-5

[117] Matsushima R, Fukao Y, Nishimura M, Hara-Nishimura I. NAI1 gene encodes a basic-helix-loop-helix-type putative transcription factor that regulates the formation of an endoplasmic reticulum-derived structure, the ER body. The Plant Cell. 2004;**16**(6):1536-1549. [Internet]. DOI: 10.1105%2Ftpc.021154

[118] Hawes C, Saint-Jore C, Martin B, Zheng HQ. ER confirmed as the location of mystery organelles in Arabidopsis plants expressing GFP! [1]. Trends in Plant Science. 2001;**6**(6):245-246. [Internet]. DOI: 10.1016%2Fs1360-1385%2801%2901980-x

[119] Hayashi M, Toriyama K, Kondo M, Hara-Nishimura I, Nishimura M. Accumulation of a fusion protein containing 2S albumin induces novel vesicles in vegetative cells of Arabidopsis. Plant & Cell Physiology. 1999;**40**(3):263-272. [Internet]. DOI: 10.1093%2Foxfordjournals.pcp.a029537

[120] Mcconn M, Creelman RA, Bell E, Mullet JE, Browse J. Jasmonate is essential for insect defense in Arabidopsis. Proceedings of the National Academy of Sciences of the United States of America. 1997;**94**(10):5473-5477. [Internet]. DOI: 10.1073%2Fpnas.94.10.5473

[121] Xie DX, Feys BF, James S, Nieto-Rostro M, Turner JG. COI1: An Arabidopsis gene required for jasmonate-regulated defense and fertility. Science (80- ). 1998;**280**(5366):1091-1094. [Internet]. DOI: 10.1126%2Fscience.280.5366.1091

Section 2

# Endoplasmic Reticulum in Disease States

Chapter 4

# Mitochondria-Endoplasmic Reticulum Interaction in Central Neurons

*Liliya Kushnireva and Eduard Korkotian*

## Abstract

The proteins presenilin-1/2 play a key role in the interactions between mitochondria and the endoplasmic reticulum at synaptic contacts of central neurons. Several novel observations suggest that mutations in presenilin-1 lead to an abnormal energy state, an early sign of neurodegeneration and Alzheimer's disease. Recent studies suggest that in the postsynaptic region, calcium stores are widely represented in the spine apparatus, which is located in a strategically important compartment - the neck of mature mushroom-shaped dendritic spines. Moreover, in the dendritic shaft area, at the base of the spines, one finds oblong mitochondrial clusters supplying the postsynaptic area and the local protein synthesis with ATP. Calcium signals, generated by the postsynaptic membranes, affect both calcium release from local stores through ryanodine channels and the uptake based on store-operated calcium entry. The entire complex of nanoscale signaling most likely determines the production of ATP. Violation of the functional relationship between mitochondria and reticular calcium depots can lead to disruption of signaling pathways that stimulate ATP production at the stages of increased activity of individual synapses. In this chapter, we will present the signaling mechanisms of interaction between mitochondria, spine clusters, and calcium nano-stores in postsynaptic area.

**Keywords:** calcium, ATP, neurodegeneration, presenilin, mitochondria, calcium store, spine apparatus

## 1. Introduction

The endoplasmic reticulum (ER) is the largest intracellular organelle in neurons. It ranges from the nuclear membrane through the axon to presynaptic terminals, and through all dendritic arbors, penetrating into some dendritic spines in the form of thin smooth tubules or their extensions such as spine apparatus [1]. Internally differentiated into smooth (sER) and rough (ribosomal, rER), the ER performs many cellular functions, including the synthesis and transport of essential intracellular molecules. However, the most enigmatic and least explored function of the ER is the storage and transmission of $Ca^{2+}$ signals from a small compartment called the spine apparatus (SA), which is located mainly in mature, mushroom-type dendritic spines [2]. This is a small multilayer lamellar structure, sometimes connected to the sER in the dendritic shaft by a thin tube passing through the neck of the spine [3]. Localization of the spine apparatus depends on synaptic activity [2]. About 80% of the large dendritic spines have a spacious SA, while only 20% of the small thin (immature) spines contain it [3] since more often the sER network can reach only the neck of the spine or be completely absent [4]. The cytoplasm of the spine has been shown to be composed of actin and actin-regulating proteins that are longitudinally located in the neck of the spine and organized into a dense lattice surrounding the sER, or SA in the head of the spine [5], whose marker is the actin-modulating protein synaptopodin (SP) [6, 7]. There are contradictions in studies involving electron microscopy (EM), which do not allow unequivocal answers to the question of whether the SA is an autonomous structure derived from the ER, indicating the final stage of the "maturity" of the spine, and all other inclusions of the ER in the spine are only transitional stages, or SA formation occurs independently of inclusion of a continuous ER compartment into the spine, and in such a case, they may sometimes coexist [3, 4, 8–12]. Nonetheless, there is a growing number of studies in which the spine apparatus is mentioned as one of the main players in the regulation of synaptic plasticity and learning and memory mechanisms. [7, 11, 13–17].

## 2. Spine apparatus functions

Calcium deposition in the postsynaptic terminal underlies synaptic transmission, driving a wide range of synaptic plasticity mechanisms for efficient learning and memory processing. Excitatory or inhibitory inputs lead to a differentiated local increase of calcium in dendritic spines, which functions as units of biophysical and biochemical computations in the neuron, regulating their duration and distribution [18]. The volume of the SA, also considered an ER store, affects the temporal dynamics of $Ca^{2+}$, its distribution to neighboring dendrite sites, and binding to numerous $Ca^{2+}$ – gated signaling pathways [2]. The volume of the spine apparatus positively correlates with the total volume of the spine [6], with the size of the spine head (which is indirectly confirmed in experiments with SP) [19] and may increase due to the activity of N-methyl-d-aspartate receptors (NMDAR) [2]. Activation of postsynaptic NMDARs by glutamate in CA1 hippocampal neurons triggers the activation of ryanodine receptors (RyR) and $Ca^{2+}$ – induced release of $Ca^{2+}$ (CICR) from the store. This release of $Ca^{2+}$ occurs and is often limited to the head of the spine [2], however, it is able to spread further along the dendrite, penetrating into adjacent spines, especially in young neurons [20]. In young (P8-P17) tissues, hippocampal postsynaptic RyRs CA3-CA1 mediate a propagating Ca2+ signal from active synapses, triggered by NMDAR-mediated $Ca^{2+}$ influx into the dendrite and adjacent coactive synapses, lowering their induction threshold for plasticity [21]. Immunological experiments show colocalization of SP and RyR in calcium stores [19]. Modeling indicates that RyRs are likely located on the SA at the base of the spine neck and their activation is triggered by the binding of two calcium ions, which move from the head to the neck inside the spine cytoplasm and generate calcium flow from the SA down into the dendritic shaft, which also is verified experimentally [22]. RyR – mediated calcium-induced calcium release (CICR) from stores can lead to either long-term potentiation (LTP) or long-term depression (LTD), depending on the pattern of synaptic activity. The mechanisms of LTP and LTD are the basis of synaptic plasticity and require an increase in postsynaptic $Ca^{2+}$ concentration ($[Ca^{2+}]i$). Caffeine, which releases

*DOI: http://dx.doi.org/10.5772/intechopen.105738*

$Ca^{2+}$ from stores, increases the number of active α-amino-3-hydroxy-5-methyl-4-isoxazolepropionic acid receptors (AMPARs) in hippocampal cultures, thus enhancing the function of synapses [19] and increasing LTP induction in hippocampal slices [23, 24]. Pharmacological blockade of CICR by RyR impairs LTP induction at hippocampal synapses [2, 11, 25]. In addition to LTP, RyRs also promote LTD, and their genetic ablation inhibits low-frequency stimulation-induced LTD in the same area of the brain [19]. In addition, it has been shown that glutamate-induced growth of new spines in the cortex is also mediated by the release of $Ca^{2+}$ from calcium stores [26].

SP-positive spines demonstrate stronger responses to glutamate uncaging than SP-negative ones. In addition, SP mediates the accumulation of the GluR1 subunit of the AMPA receptor in the heads of spines [19]. Adult mice have more mature spines containing SA and a larger area of SP inclusions than the young. Hippocampal neurons in SP knockout (SPKO) young mice lack SA and decrease LTP, do not solve cognitive tasks, and lose spatial memory. However, with age, SPKO mice develop compensatory mechanisms for LTP recovery. They demonstrate increased excitability and expression of an activity-regulated cytoskeleton-associated protein (Arc), encoded by a member of the immediate-early gene (IEG) family *ARC*, in hippocampal dentate gyrus (DG) granule cells after exposure to a novel environment. Arc mRNA and protein are known to accumulate in dendritic regions that receive high-frequency synaptic inputs [27]. Learning or stress-induced Arc accumulation in dendrites is critical for plasticity and memory consolidation [28]. DG is an area required for spatial learning processes, which in SPKO mice activates more cells than in mice with a normal genotype. Since the intracellular concentration of $Ca^{2+}$ regulates excitability through the activation of calcium-gated $K^+$ – currents, a decrease in calcium availability as a result of the absence of SP and calcium stores can cause increased excitability in the SPKO brain. Therefore, it is important to compensate for the decrease in $[Ca^{2+}]i$ in postsynaptic sites for synaptic plasticity [29].

In addition to RyR, the release of calcium from ER stores is enhanced by activation of inositol triphosphate receptors (IP3R), which are enriched in dendritic branches [19]. Activation of type I metabotropic glutamate receptors (mGluR) on the plasma membrane (PM) causes an increase in the concentration of inositol triphosphate via phospholipase C (PLC). This, in turn, enhances the release of $Ca^{2+}$ from the stores into the cytoplasm by stimulating the IP3R on the ER membrane. IP3R activation can propagate the calcium wave along the dendritic segment or be limited to postsynaptic microdomains, depending on the ambient level of inositol triphosphate in the cytoplasm. IP3R is able to be co-activated by intracellular $Ca^{2+}$ and these two mechanisms, RyR-mediated CICR and IP3R-induced release of calcium from the stores, are capable of mutual reinforcement [30]. Thus, the SP -associated $Ca^{2+}$ – stores in the form of a SA play an important role in the storage and regulation of $Ca^{2+}$ secretions needed for neuronal plasticity.

In ER stores, $Ca^{2+}$ is bound to calcium-binding proteins (CBPs), such as calnexin and calreticulin. Each CBP binds to many Ca2+ ions in a low affinity and high-capacity manner. When the $Ca^{2+}$ store is open, exporters can easily separate $Ca^{2+}$ from the CBP. The store also maintains the concentration of free $Ca^{2+}$, which determines the driving force for the release of $Ca^{2+}$. To compensate for the depletion of the $Ca^{2+}$ pool, its accumulation in the ER is carried out through the use of $Ca^{2+}$ pumps of the $Ca^{2+}$-ATPase family (SERCA) together with a store-operated calcium entry (SOCE) via store-operated calcium channels (SOCs). The functioning of SOCs depends on the $Ca^{2+}$ concentration inside the store and calreticulin, which accounts for nearly half of total ER $Ca^{2+}$ binding, acts not only as a $Ca^{2+}$ buffer but also as an important chaperone and regulator of SERCA pumps. Calcium depletion from ER stores is determined by stromal interaction molecules (STIMs), which are diffusely distributed throughout the resting ER. When the storage is empty, STIM transmembrane calcium sensors accumulate on the membrane of the depleted ER store, closest to PM, where they activate the voltage-independent calcium release-activated protein (Orai), providing an influx of $Ca^{2+}$ into the store so as to replenish it (**Figure 1**) [30]. Two homologs of STIM, STIM1 and STIM2, are found in neural tissue but appear to be associated with different functions in developing and mature neurons, although both are associated with the Orai channel. STIM1 clusters predominate in young cells, they are more mobile, and their movement along dendrites triggers local $Ca^{2+}$ transitions, while STIM2 clusters are active in mature neurons, are more dispersed, and much less mobile. STIM1 plays an important role in the formation and functional maturation of filopodia and growth cones in young cells, and STIM2 binds to SOC under conditions of $Ca^{2+}$ deficiency, restoring local $[Ca^{2+}]$i levels and moving into active dendritic spines [31]. Inhibition of SERCA pumps by thapsigargin results in slow $Ca^{2+}$ leak from the ER, stimulating STIM1 oligomerization and formation of STIM1/Orai1 complexes, and antagonists of STIM/Orai dependent SOCE reduce LTP in hippocampal neurons. Synaptic activity is critical for maintaining the morphology of mature dendritic spines, as blockade of synaptic activity by tetrodotoxin (TTX) causes STIM-associated increases in spontaneous calcium transients and a decrease in the proportion of mature spines versus the proportion of immature filopodium [31, 32].

The STIM/Orai complex is also studied in the context of the development of neurodegenerative diseases, in particular Alzheimer's disease (AD). STIM1 and STIM2 are involved in maintaining $Ca^{2+}$ homeostasis in neurons and are involved in the production of beta-amyloid peptide (Aβ), which accumulates in AD. Overexpression of these proteins can initiate pathological activation or deactivation of SOCE-dependent mechanisms, disrupting synaptic transmission and thus stimulating neurodegenerative mechanisms [31, 33].

## 3. Structural plasticity of spines

LTP and LTD are associated with an increase and decrease in spine volume, respectively. Similar to functional plasticity, structural plasticity also requires Ca2+ influx through postsynaptic NMDARs, activation of $Ca^{2+}$/calmodulin-dependent protein kinase II (CaMKII) for recruiting GluR subunits, small guanosine triphosphatases (GTPase) and actin polymerization [34]. What specific interactions are required in local targeting of mRNA and protein synthesis to increase the morphology of active spines? To answer this question, the synaptic tagging and capture (STC) hypothesis was proposed: LTP induction creates a "tag" in potentiated synapses that can capture plasticity-related proteins (PRPs), including Homer protein homolog 1a (Homer1a) and Arc. Homer1a, a postsynaptic scaffold protein, is recruited from the soma into the stimulated spine. A synaptic tag can be a temporary morpho-functional state of the synapse, which is represented by a complex of proteins in interaction with the structures of the actin cytoskeleton. For example, LTP is known to induce the formation of a stable pool of F-actin that can act as a synaptic tag. However, this tag is also found in unstimulated spines. So, after LTP, Arc accumulates in unstimulated spines and is excluded from potentiated ones. The amount of synaptic Arc is negatively correlated with the amount of surface GluA1 at the synapses and promotes AMPAR endocytosis. It is likely that this reverse synaptic labeling helps maintain the synaptic weight contrast between active and inactive spines in areas of high synaptic plasticity such as the hippocampus [27, 34]. This mechanism is especially attractive when the spines are close to each other and are part of the same functional synaptic cluster. It is known that morpho-functional changes correlate among neighboring spines. Decrease in the LTP induction threshold decays along ~10 μm of the dendritic branch, and in young neurons, the repeated release of glutamate from individual spikes reduces the induction threshold in the surrounding area to several minutes [34]. We hypothesize that intra-cluster differentiation of synapses is essential for structural plasticity. For example, after LTP induction, the activity of GTPases Ras and Rho extends to ~5–10 μm along the dendrite with the ability to penetrate into neighboring spines. In addition, single spike activity can also trigger molecular signaling involving calcineurin, IP3R, and metabotropic glutamate receptor 1 (mGluR1), which act on adjacent spines [35]. It is possible that Arc-dependent depotentiation of non-target spines helps de-tag neighboring spines for accurate PRP capture.

The mechanism linking local synaptic events in individual spines and signals entering the nucleus includes such transcriptional regulators as cAMP response element-binding protein (CREB) and synapto-nuclear protein messenger *Jacob*, which are translocated into the nucleus in response to synaptic activity. Upon NMDAR activation, *Jacob* is phosphorylated and dissociated from the spines, translocating to the nucleus in an importin-dependent manner. The presence of phosphorylated *Jacob* in the nucleus increases CREB phosphorylation, inducing the expression of CRE-dependent genes that are involved in the generation of the protein tag only in potentiated synapses [34]. However, such a tag is likely not so much a complex of protein molecules as a short-term reconfiguration of the spine cytoskeleton due to CaMKII activity, including restructuring of postsynaptic density (PSD) and the number of AMPAR. During LTP, the autophosphorylated form of CaMKII remains active in the PSD even after the $Ca^{2+}$ concentration returns to baseline levels and such a tagged synapse is able to capture PRP. Such synapse stabilizes its new structural conformation before the tagging state disappears, and thus will retain a change in its synaptic efficiency. Inhibition of CaMKII autophosphorylation disrupts the tag, interfering with structural, but not early functional plasticity [36].

If we also turn to SA as a potential tag of functionally improved synapses, like Ostroff et al., who have shown that the presence of SA in large spines correlates with the presence of polyribosomes in their heads after learning. This may be an indicator of a high degree of spatial specificity of translation in dendrites. Accumulation of polyribosomes in the heads of large spines reflects the enhancement of translation during learning, suggesting a link between structural plasticity and memory consolidation [37].

## 4. The role of mitochondria

The development, maturation, and maintenance of functioning synapses require the maintenance of ionic balance, proteostasis, and modification of synaptic proteomes at the expense of a significant amount of energy [38]. Of great importance is the local specificity of these processes for plasticity, which has an advantage in comparison with the generalized delivery of proteins from the soma. To meet global and local energy needs, cells regulate mitochondrial movement, fission, fusion, and "parking" mitochondria in every area of the cell. Neuronal mitochondria play a crucial role in maintaining synaptic functions, in particular, by providing local translation both under basic conditions and during plasticity processes [38–40]. Rangaraju et.al from Schuman's laboratory used adenosine triphosphate (ATP) inhibitors, local inhibition of mitochondrial compartments, and visualization of newly synthesized proteins, to show that local translation is provided by local mitochondrial clusters. Translation of proteins necessary for morphological plasticity, going on from mRNAs, localized near synapses. Glutamate-inducted LTP was not established in areas containing nonfunctioning mitochondria, preventing the morphological changes characteristic of this mechanism. Also, in mitochondria-free regions, there was a significant decrease in protein synthesis compared to stimulated regions with functional mitochondria, although at basal levels of neuronal activity, the energy needs of local translation are adequately met by global levels of ATP available in the dendrite or ATP produced by neighboring functioning mitochondria [38].

Fluctuations in cytosolic calcium, glucose and ATP levels, synaptic activity, neurotransmitters, and growth factors are able to regulate positioning, mitochondrial transport, and dynamics. Moving these organelles to energy-demanding sites, such as synapses, dendritic spines and axons [41], is essential for their functioning. Mitochondrial morphology differentiates not only depending on cell type but also on localization in a particular cellular compartment and can change with age [42]. *In vitro* and *in vivo* imaging of axonal mitochondria often shows them to be short, while dendritic mitochondria have a more elongated morphology and often show greater complexity than axonal [43]. This complexity can be expressed at least in terms of length or volume and tendency to form clusters of mitochondria in the form of tubes, often overlapping with each other in dendritic shafts. For example, in the experiments of Faitg and others, the average volume of individual mitochondria in hippocampal neurons varied between 0.11 and 1 $\mu m^3$, while axonal mitochondria had an average volume of 0.12–0.27 $\mu m^3$ and did not reach values >1 $\mu m^3$; somatic mitochondria occupied intermediate values (0.14–0.16 $\mu m^3$) [42]. In another study with cortical neurons, the length of dendritic mitochondria varied from 0.52 to 13.28 μm, while the length of axonal mitochondria was much smaller from 0.3 to 1.13 μm [43]. More than 50% of dendritic extensions are filled with mitochondria, but less than in axons 1% are filled [38, 43]. Another study compared mitochondria in the presynaptic bouton and in the postsynaptic dendrites of the hippocampus. The area of presynaptic mitochondria was almost half, on average, 0.077 $\mu m^2$ versus postsynaptic 0.146 $\mu m^2$. In addition, presynaptic mitochondria were significantly darker, i.e., had a higher electron density than postsynaptic [44]. It is possible that the elongated morphology of mitochondria in dendrites provides them with a bioenergetic advantage [45], while an increase in density indicates their greatest activity. Tubular mitochondrial stretches in dendrites are observed in hippocampal neurons, the average length of which sometimes even exceeds 10 – 30um in length [38, 46]. These may be called a mitochondrial cluster, since such a structure more likely represents individual mitochondria undergoing longitudinal outer (OMM) and inner mitochondrial membranes (IMM) fusion [47, 48], because normal elliptical mitochondria are smaller under normal conditions (see numbers above). Such clusters can be cleaved by mitochondrial uncouplers such as FCCP or CCCP [49]. Clustering of mitochondria increases their functional stability, optimizing the use of the local mitochondrial pool in cell compartments under basic conditions and during activity. Fusion and fission, the direction of movement and parking of mitochondria in certain areas of neuronal processes are of great importance for meeting the changing energy needs of various areas of the cell.

## 5. Mitochondrial movement and parking

In neurons, the movement of mitochondria is realized by their attachment to microtubules and it depends on their polarity [50]. In axons, microtubules have a distinct polarity, so that anterograde transport of mitochondria along (+)-terminal microtubules move them toward the growth cone or presynaptic end of the axon using kinesin family proteins (KIFs), while retrograde transport along (−)-terminal microtubules use motor protein complex of dynein, directing mitochondria toward the soma. In dendrites, microtubules can show mixed polarity so that KIFs and dynein motors can drive cargo transport in dendrites either anterograde or retrograde depending on microtubule polarity [50]. There are also reports that the movement of mitochondria over short distances in areas rich in actin cytoskeletons, such as the axon growth cone, presynaptic terminal, and sometimes in large dendritic spines observed in the cortex and hippocampus, is carried out with the help of myosin motors [46, 50]. The share of movement mitochondria varies from 5 to 20 to 35–45% of the mitochondrial pool in a culture of hippocampal neurons, and mitochondrial movement occurs more intensely in axons than in dendrites, where mitochondria can often be relatively immobile both in synaptic and non-synaptic areas. Movement increases during blocking of activity by TTX in all outgrowths, while induced glutamate excitotoxicity induces a persistent increase in $[Ca^{2+}]$i, slowing down movement and promoting mitochondrial rounding in all neuronal outgrowths [45]. Under physiological conditions, $Ca^{2+}$ influx occurs in areas of high metabolic demand, such as axon terminals and complexes of postsynaptic structures, where mitochondria tend to accumulate. According to various data, from 36% to ~50% of presynaptic axon terminals of hippocampal neurons contain mitochondria [38, 45]. Sustained elevation of $[Ca^{2+}]$ i levels in some areas of the cell may be a marker of local activity or ATP deficiency, which will recruit and retain mitochondria to stop in these areas, thus balancing ATP production with local needs [51]. Numerous signaling pathways have been elucidated that are regulated by $Ca^{2+}$, modulating mobility or inducing arrest of mitochondrial pools. Mitochondrial microtubule-based mobility is mediated by KIF $Ca^{2+}$-dependent manner, and the OMM-anchored mitochondrial Rho GTPase 1 (Miro1) is a $Ca^{2+}$ sensor (because it contains two helix-loop-helix EF-hand $Ca^{2+}$-binding motifs) for mitochondrial localization in synapses, and its association with PTEN-induced kinase 1 (PINK1) and recruitment of the Parkin protein is necessary for mitochondrial motility stops [50, 52, 53]. Mitochondrial arrest during neuronal activity is produced by synapse-released glutamate, which activates NMDA receptors and induces $Ca^{2+}$ – influx, which binds to Miro1. The miro1-mediated mitochondrial arrest may recruit passing mitochondria to active synapses where ATP and calcium buffering requirements would be higher [53]. In a culture of rat hippocampal neurons, perfusion of 30 μM glutamate (with 1 μM glycine to activate NMDA receptors) for 10 min reduced the number of moving mitochondria by 95% in the presence of extracellular calcium ($[Ca^{2+}]$o), but only a 28% decrease in movement was observed with absence of $[Ca^{2+}]$o. In the same study, it was demonstrated that moving mitochondria were stopped in areas that were positive for the synaptic marker synaptophysin and synaptic vesicle protein 2 (SV2), and the distribution of mitochondria in dendrites to the nearest synapse was reduced after the application of glutamate. All of this indicates that mitochondria are recruited to synaptic zones during activity [53]. In addition to these mechanisms, a positive correlation is also found between anterograde mitochondrial transport in axons and electrical potential through IMM (MtMP, $\Psi$m), increasing ATP production. Thus, the maximum rate of ATP production in isolated mitochondria approximately doubles with an increase in Ψm for every 10 mV [54]. Mitochondria with high membrane potential also tend to accumulate at synapses [55]. It can be concluded that the motility and spatial distribution of mitochondria for ATP production can be dynamically regulated by local [$Ca^{2+}$]i.

## 6. The role of mitochondria in $Ca^{2+}$ buffering and signaling

Mitochondria play an important role in $Ca^{2+}$ signaling in neurons. They have a buffering capacity for Ca2+ binding that arises from their highly hyperpolarized membrane. Membrane potential values of functioning mitochondria in a culture of rat cortical neurons vary within ≃100 – ≃160 mV [54]. MtMP allows rapid transfer of cytosolic $Ca^{2+}$ across the IMM along its electrochemical gradient into the organelle by the mitochondrial calcium uniporter (MCU), despite the last low affinity for $Ca^{2+}$ ions. The mitochondrial matrix accumulates high levels of $Ca^{2+}$ both during global $Ca^{2+}$ signals and in response to a moderate local increase in [$Ca^{2+}$]i [56]. This uptake is balanced by $Ca^{2+}$ efflux through the $Na^{+}$/$Ca^{2+}$ antiporter or, more rarely, through the mitochondrial permeability transition pore (mPTP), these pathways are also sensitive to mitochondrial membrane depolarization [56, 57]. Small transient mitochondrial depolarizations reflect mitochondrial buffering activity in high [$Ca^{2+}$]i micro-domains. From studies in isolated mitochondria, uptake of 17 μM $Ca^{2+}$ caused a 2–3 mV mitochondrial depolarization [58], while a much greater loss of MtMP accompanies opening of mPTP, nonselective pores in the inner mitochondrial membrane. Through these any molecule weighing less than 1500 da can penetrate, which can lead to an increase in osmotic pressure, swelling of mitochondria and subsequent rupture of the OMM, or cause depletion of matrix substances, incl. and $Ca^{2+}$ [55]. Temporary opening of the pore can be caused by $Ca^{2+}$ overload and is characterized by uncoupling of the oxidative phosphorylation chain, resulting in a decrease in ATP production, but is reversible upon restoration of cellular homeostasis. In contrast, the permanent opening of mPTP triggers mitochondria-mediated apoptosis due to the release of cytochrome C [30, 57]. Mitochondria are able to buffer about 75 μM of $Ca^{2+}$ before it is released, and this point of maximum buffering seems to depend on the rate of $Ca^{2+}$ uptake. Rapid mitochondrial depolarization also leads to subsequent calcium release, as shown in experiments with CCCP [59]. Interestingly, the content of $Ca^{2+}$ in mitochondria is inversely proportional to the rate of mitochondrial movement. Motile mitochondria tend to have a lower intra-mitochondrial $Ca^{2+}$ signal ([$Ca^{2+}$]m), however, the [$Ca^{2+}$]m levels do not affect the direction of movement, and no difference was shown in [$Ca^{2+}$]m between anterograde and retrograde mitochondrial movement. We have previously said that microtubule-based mitochondrial motility is driven by [$Ca^{2+}$]i levels, however direct $Ca^{2+}$ influx into the mitochondrial matrix via the Miro1 mediated MCU is also capable of altering mitochondrial motility. In experiments with mutations in the EF-hand domains of Miro1, mitochondrial movement did not stop, despite an increase in [$Ca^{2+}$]i levels. Blocking the MCU also allowed mitochondrial movement to be preserved even in the presence of high levels of cytoplasmic [$Ca^{2+}$]i, but only partially, indicating that [$Ca^{2+}$]i also makes a significant contribution to mitochondrial mobility. Miro1 can change the level of $Ca^{2+}$ influx into mitochondria, acting as a [$Ca^{2+}$]i sensor, similar to STIM ER, influencing the amount of $Ca^{2+}$ influx into mitochondria and their rate, which can be regulated by the amount of Miro1 protein associated with kinesin motors at a given time, while as soon as the concentration of [$Ca^{2+}$]m reaches a critical level, the interaction of Miro1 with kinesin complexes is disintegrated and mitochondria stop. The influx of $Ca^{2+}$ into mitochondria increases ATP production by activating the tricarboxylic acid cycle, as well as increasing the activity of electron transport chain enzymes and the ATP synthase complex, so that ultimately mitochondrial arrest is beneficial near synaptic areas [60].

## 7. ER-mitochondria interactions

Mitochondrial clusters occupy most of the dendritic branches and are closely associated with the ER, especially at the base of the spines (**Figure 1**) [18]. However, such a distribution of ER and mitochondria is not constant. Contacts between mitochondria and the ER along the dendrites enable functional inter-organellar communication and play a central role in the regulation of postsynaptic calcium-signaling, and dysregulation of its communication has been demonstrated in neurodegenerative diseases such as Alzheimer's and Parkinson's diseases [18]. Contact sites between OMM mitochondria and ERs (Mitochondria-ER Contact Sites, or MERCs), constituting about 2–20% of the mitochondrial surface area, were detected using EM studies, then demonstrated in experiments using dimerization-dependent fluorescent proteins. Membrane constituents from a specific set of protein and lipid complexes that form MERC are called mitochondria-associated ER membranes (MAMs). The term MAM is used to describe the results of experiments on the biochemical/functional characterization of isolated contacts between mitochondria and the ER, while the term MERC is used during morpho-functional visualization [61].

MERCs provide a direct route for calcium transfer from the ER to the mitochondria and are required for mitochondrial functions, including ATP production. An increase in MERC surface area increases mitochondrial calcium influx and hence stimulates ATP production [18]. The MERC width is not constitutive and can change depending on the metabolic status of the cell. Excessive expansion, or vice versa, constriction of MERC disrupts the efficient transfer of calcium from the ER to the mitochondria. $Ca^{2+}$ uptake by the MCU is most effective when the ER membrane faces the OMM at a distance of ≈15 nm, but not more than 30 nm, when calcium will leak and diffuse into the cytoplasm and not less than 10 nm, which is determined by the length of the protruding part of IP3R or RyR [61]. The $Ca^{2+}$ release from the ER is able to control mitochondrial mobility by locally increasing [$Ca^{2+}$]i. Inhibition of movement of mitochondria at an optimal distance from active $Ca^{2+}$ receptor channels from the ER is necessary for mitochondrial calcium buffering, which also serves as a means of stimulating mitochondrial ATP production. Although mitochondria store less $Ca^{2+}$ than ER, they are important buffers of [$Ca^{2+}$]i. Voltage-gated anion channels (VDACs) located on the OMM are responsible for the rapid transfer of $Ca^{2+}$ from the ER to mitochondria, and their function results in the formation of $Ca^{2+}$-rich microdomains in the mitochondrial intermembrane space [62]. Balance of [$Ca^{2+}$]m is maintained by influx through the MCU and outflow through the mitochondrial sodium-calcium exchanger (NCLX), but at elevated levels of activity, mitochondria are also able to buffer excess [$Ca^{2+}$]i due to precipitation inside the matrix in the form of $Ca^{2+}$ – phosphate [30]. Sigma-1 receptors (σ-1R) found on MAM may influence mitochondrial $Ca^{2+}$ – signaling. σ-1Rs are able to exhibit chaperone activity, stabilizing IP3R to MAM under conditions when $Ca^{2+}$ reserves in the ER are depleted, preventing degradation of $Ca^{2+}$ entry into mitochondria, restoring Ψm and subsequent ATP production (**Figure 1**) [62].

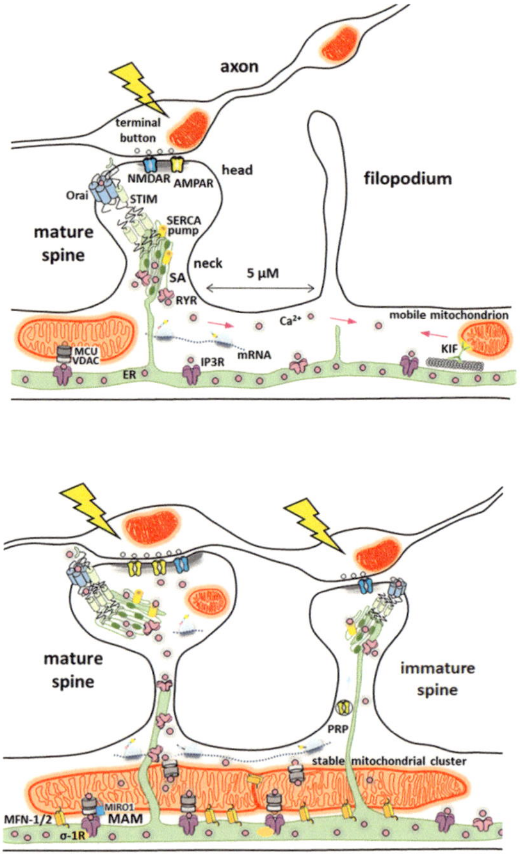


**Figure 1.**

*Top. Mature mushroom-shaped dendritic spine contains spine apparatus (SA), which is required for synaptic transmission and plasticity. SERCA pump mediates $Ca^{2+}$ uptake into SA due to synaptic activity or STIM/ Orai-dependent store-operated calcium entry mechanism. Excitatory input (lightning) activates calcium influx through postsynaptic NMDAR, which triggers calcium-induced calcium release from SA through RyR. This $Ca^{2+}$ transient spreads along the spine neck toward the dendrite. In turn, this initial transient trigger molecular signaling leads to the release of $Ca^{2+}$ from $IP_3Rs$, which are localized on the dendrite-running ER. Calcium released from RyR and $IP_3R$ may attract Mobile extrasynaptic mitochondria located within 5–10 μm to the base of the spine. Mobile mitochondria move along microtubules driven by a protein of the kinesin family (KIF). Continuous $Ca^{2+}$ transients at the base of mature spine dock the mitochondrion by dissociating it from microtubule. Retention and accumulation of mitochondria take place in the area of locally elevated calcium around ER. Calcium originated from $IP_3R$ covers an area around voltage-gated anion channels (VDAC), the intermembrane space of mitochondria, and then through the mitochondrial calcium uniporter (MCU) moves to the mitochondrial matrix. Stable postsynaptic mitochondria provide ATP-dependent mechanisms of plasticity, the local ribosome-dependent translation of proteins on the base of mRNAs possibly attracted by local $Ca^{2+}$ gradients. The nearby filopodium does not receive presynaptic inputs and is therefore unable to take part in synaptic plasticity. Bottom. Mature and immature dendritic spines both receive excitatory presynaptic inputs. A more intense postsynaptic response occurs in a mature spine containing both NMDA and AMPA receptors. Synaptic input initiates calcium release from the SA and ER, inducing local translation of plasticity-related proteins (PRPs). PRPs can also penetrate into the adjacent, co-activated but yet immature spine, causing the incorporation of AMPA receptors. A stable postsynaptic mitochondrial cluster sequesters $[Ca^{2+}]i$, which has been released during activity. This mechanism highly limits calcium distribution along the dendrite, maintaining individual spine autonomy. New mitochondria fuse into a cluster. Their contacts with the ER are maintained by mitofusins (Mfn) 1/2. Sigma-1 receptor (σ-1R) helps to stabilize and attach $IP_3R$ to VDAC in mitochondria-associated ER membranes (MAM). $Ca^{2+}$ influx through MCU is also mediated by mitochondrial rho GTPase 1 (Miro1), which is able to function as a $[Ca^{2+}]m$ sensor. Miro1 modulates the level of $Ca^{2+}$ influx into the mitochondrial matrix. It is also involved in MAM - mitochondria stabilization mechanism. The larger volume of the mitochondrial cluster enables the production of higher levels of ATP for efficient PRP production, promoting functional and morphological plasticity. Individual mitochondria are also shown to be able to enter the heads of the mature spines with large head volume and SA, where they possibly feed local translation mechanisms.*

Knockout of Sig-1R in hippocampal neurons results in shorter and smaller mitochondria and also causes a decrease in MtMP and release of cytochrome c, which leads to disruption of cytoskeletal networks loss of mature dendritic spines, and formation of immature dendritic spines [63, 64]. Genetic deletion of σ-1R also impairs MAM stability and leads to a decreased number of MERC [65]. A reduced amount of σ-1R has been observed postmortem in the hippocampi of AD patients, and certain σ-1R gene polymorphisms coexist with the well-known risk factor Alzheimer's disease apolipoprotein ε4 (APOE ε4) [66].

Mitochondrial fission and fusion can also be regulated by the ER, for example, by Wnt-5a, a member of the Wingless/integrase (Wnt) secreted glycoprotein family, which induces an increase in $Ca^{2+}$ efflux from the ER via IP3R and RYR. This, in turn, activates $Ca^{2+}$-dependent signaling molecules, including CaMKII, protein kinase C (PKC), and calcineurin, which may promote phosphorylation of dynamin-related protein 1 (Drp1) associated with increased mitochondrial fragmentation [41]. Although at rest, fragmented dendritic mitochondria are less beneficial for energy production, but they are able to increase ATP production at high levels of arousal through higher $H_2O_2$ production [67].

## 8. Functionally and spatially related clusters of spines are served by mitochondria clusters

Modeling and electrophysiological experiments indicate that neighboring synaptic inputs can sum nonlinearly and turn a dendritic compartment or a cluster of spines into a separate summing unit of activity. The experimentally observed signs of synaptic clustering are expressed in the sharing of synaptic inputs from the same axon by two or more neighboring spines. That is, the formation of synapses and their clustering is a consequence of their relative position with the nearest axon (**Figure 2**) [35]. The presence of clustering is also evidenced by the above mechanisms of transfer of intracellular signaling molecules by diffusion from the stimulated spine to a certain distance to neighboring dendritic spines, which can "learn" a similar pattern of activity. There is also evidence for intercluster competition: LTP induction in multiple spines on the same dendritic segment can cause spine contraction and weakening of the synapses of neighboring, more distant, unstimulated spines [34].

It can be expected that the united cluster of spines should have its own stable energy source in the form of a mitochondrial cluster, which will limit $Ca^{2+}$ diffusion and ATP production mainly to this region. The calculations by Rangaraju et al. suggest that mitochondrial clusters spanning about 30 μm of dendritic length are capable of local ATP synthesis to provide energy for 30–300 spines (with a uniform distribution of 1–10 spines per μm of dendrite length) [38]. This co-compartmentalization provides an advantage at the time of obtaining simultaneous excitation for several dendritic spines, in which ATP production from one local mitochondrion or ATP levels in the dendrite would be clearly insufficient, and the expectation of ATP influx from neighboring mitochondria would adversely affect time-dependent plasticity and competitiveness given synaptic cluster. High-frequency excitatory input that leads to a significant increase in the local level of $[Ca^{2+}]i$ also require immediate absorption, which is successfully performed in the case of mitochondrial clustering. Could be supplemented to reformulate Hebbian plasticity into: "Spines that are repeatedly active at the same time with the same inputs will tend to become 'associated' so that activity in one will facilitate activity in the other. Together, they represent a cluster unit of information in the brain and it will be provided by the own mitochondrial cluster."

What are the advantages of such a clustered mitochondrial service structure for a single spine? Modeling shows that ATP availability in dendrites is directly proportional to mitochondrial length [18]. In the case when several spines of the cluster simultaneously receive excitation, it will cause a more extensive total postsynaptic response, and then such a cluster, which has an extended mitochondrial cluster under it, acting, in this case, as a whole, will be provided with more energy. In pathological cases, the breakdown of such a cluster organization can occur: a decrease in the number of mature functional spines, which will lead to disorganization of the synaptic cluster, and a violation of mitochondrial recruitment and mitochondrial clustering will make energy production more chaotic.

Will an individual spine that receives a local signal be "lost" in such a unified structure? It is known that within a cluster, spines may compete for limited resources (**Figure 2**). After LTP induction, there is a marked loss of small spines, which is accompanied by an increase in the remaining spines, so that the total synaptic surface area per dendrite length remains constant [35]. There are hypotheses, the essence of which is that, despite the apparent morphological integrity of the mitochondrial cluster, the limited spaces represented by the cristae compounds create lateral gradients of critical metabolites and macromolecules within the cristae. Thus, in the mitochondria of the brain, the cristae are connected to the surface of the IMM through narrow long tubular sections. Modeling suggests that such compounds may lead to micro-compartmentation within the mitochondria, which may have important functional implications by creating a barrier to the diffusion of molecules between the cristae and intermembrane spaces. Rearrangements of such barriers change the energy output [68], allowing local ATP targeting.

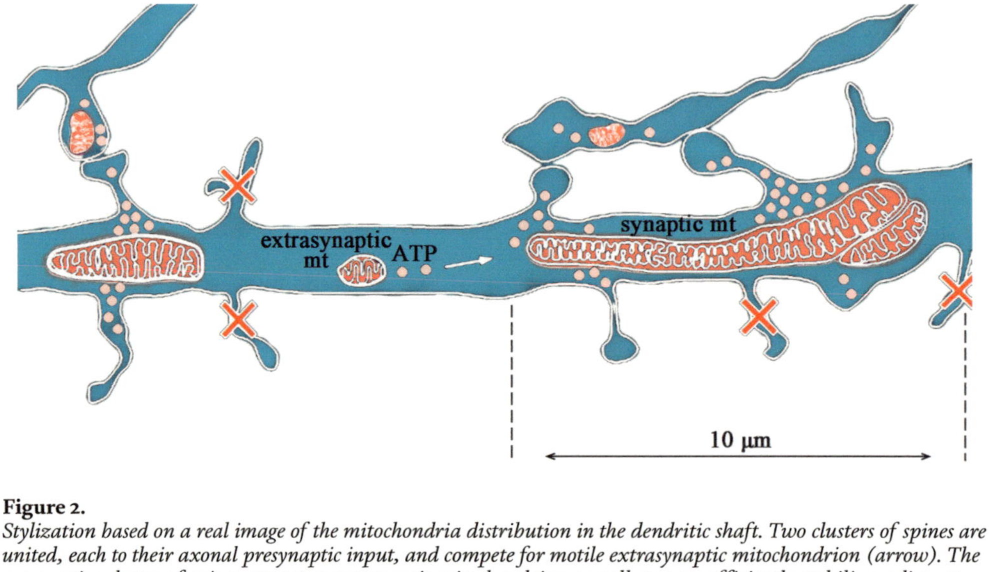


**Figure 2.**
*Stylization based on a real image of the mitochondria distribution in the dendritic shaft. Two clusters of spines are united, each to their axonal presynaptic input, and compete for motile extrasynaptic mitochondrion (arrow). The more active cluster of spines attracts extrasynaptic mitochondrion as well as more efficiently stabilizes adjacent mitochondrial cluster due to an increase in ATP demand. Active spines receive higher levels of ATP from their local synaptic mitochondrial, compared to spines that do not receive sufficient synaptic inputs. Therefore, both intercluster and intracluster spine competition can be proposed.*

## 9. The role of presenilin-1 in interactions between mitochondria and the ER in the neurodegeneration context

The “MAM hypothesis”, for AD suggests that this disease is a consequence of a disruption of ER-mitochondrial interactions, since studies show an increased expression of MAM-associated proteins in the brain of humans and mice with AD, up to the appearance of Aβ plaques, which precedes the more common “amyloid hypothesis”. For example, fibroblasts obtained from AD patients whose symptoms include impaired lipid metabolism have elongated, up to more than 200 nm MERC and increased lipid traffic [61]. PS1 and PS2 are two highly homologous isoforms of mammalian presenilin, with mutations associated with about 40% of all known cases of familial AD. Their gamma-secretase activity is enriched in MAM where they affect the processing of amyloid precursor protein (APP) to form Aβ. PS-1 plays a key role in the interactions between mitochondria and the ER in the area of synaptic contacts. Presenilin-2 (PS2) in the presence of the mitochondrial fusion protein mitofusin 2 (Mfn-2) located in OMM associated with mitofusin 1 (Mfn-1) on the ER membrane promotes mitochondrial binding to the ER [30]. Mutant presenilins have also been implicated in disrupting $Ca^{2+}$ signaling in neurons due to the release of excess amounts of $Ca^{2+}$ from the stores with the help of RyR and IP3R. RyR expression levels are elevated in cultured neurons expressing mutant PS1 in 3xTg-AD mice starting at a young age [69].

The most interesting and not related to the gamma-secretase activity of presenilins is their participation in the role of channels of passive leakage of $Ca^{2+}$ ER in the hippocampus, shown in the laboratory of Bezprozvanny [70]. The increase in resting $Ca^{2+}$ levels observed in PS1-transfected cells may be due to “leak” storage and the fact that the SERCA pump takes longer to pump the leaked $Ca^{2+}$ back to the store. Calculations predict that under physiological conditions the SERCA pump reaches thermodynamic equilibrium when $[Ca^{2+}]$ER is 2.4 mM, however, visualization gives different values, in the range of 100–500 μM $[Ca^{2+}]$ER. The researchers proposed an explanation for this difference as the presenilin-mediated leak of the ER membrane to $Ca^{2+}$ ions. Then the stationary intraluminal level of $Ca^{2+}$ in the ER is determined by the balance between its injection with SERCA and passive leak into the cytosol. However, the PS1-M146V mutant was not able to function as a $Ca^{2+}$ leak channel, which led to an overflow of $Ca^{2+}$ stores and an increase in the level of released $Ca^{2+}$ upon IP3R activation [70].

According to an alternative hypothesis, presenilins do not necessarily directly regulate the leakage of calcium ions from cisterns in the endoplasmic reticulum. It is possible that they serve as regulators of the family of ryanodine receptors (RyR1–3), which in turn are responsible for the leakage due to CICR [71]. Several studies have established close proximity and molecular linkage between these proteins on the ER membrane [71, 72]. The deep functional relationship between presenilins and RyR is highlighted by the fact that violation of control or mutation of PSs leads to increased expression of RyR2 and RyR3 [71]. Increased levels of RyR expression are found in models expressing PS1 mutants, possibly as a compensatory mechanism associated with loss of PS leakage function [71]. It has been also suggested that N-termini fragments of presenilin-1 or -2 interact with RyR, significantly increasing its sensitivity to cytosolic calcium and the opening probability of this channel (**Figure 3**) [73]. Mutant presenilin can drastically reduce the RyR gating rate and, consequently, the leakage of calcium ions from the ER depo. However, it is important to mention that RyR leakage is directly dependent on cytosolic calcium levels. Consequently, the leak will be enhanced at high synaptic activity and in those locations on the dendritic shaft where synapses are located. In numerous studies, it is noted that the production of ATP by mitochondria is closely related to the release of calcium through RyR in the MAM region [74, 75]. It is likely that a drop of calcium release through RyR in the MAM zone due to presenilin mutation does not provide the mitochondria with a sufficient calcium signal to initiate ATP production, even in areas of high synaptic activity. Moreover, PS mutations have been found to overload the ER with calcium ions, which are "locked" inside due to reduced leakage in mutant PSs, particularly in Alzheimer's disease (**Figure 3**) [76, 77].

Postsynaptic areas are among the major consumers of energy, which is used to maintain ionic balance, enzymatic processes, and local protein synthesis. Synapses are associated with large and sharp fluctuations in the levels of calcium ions, which are involved in signaling at the junction of the cytosol, ER, and mitochondria. This signaling is aimed at rapid and fine regulation of ATP levels. The energy supply and trafficking of mitochondrial clusters are vital for maintaining synaptic transmission. Thus, in a study by Du et al. [78], it was found that impairment of functionality, permeability, and trafficking of synaptic mitochondria occurs much earlier than among extrasynaptic mitochondria as well as long before the generation of amyloid plaques in a mouse model of AD. Histopathological hallmarks that occur during brain aging or AD are associated with functional insufficiency of synaptic mitochondria. Damaged mitochondria gradually increase and generalize oxidative stress, synaptic dysfunction, loss of contact, and culminate in neuronal degeneration [79, 80].

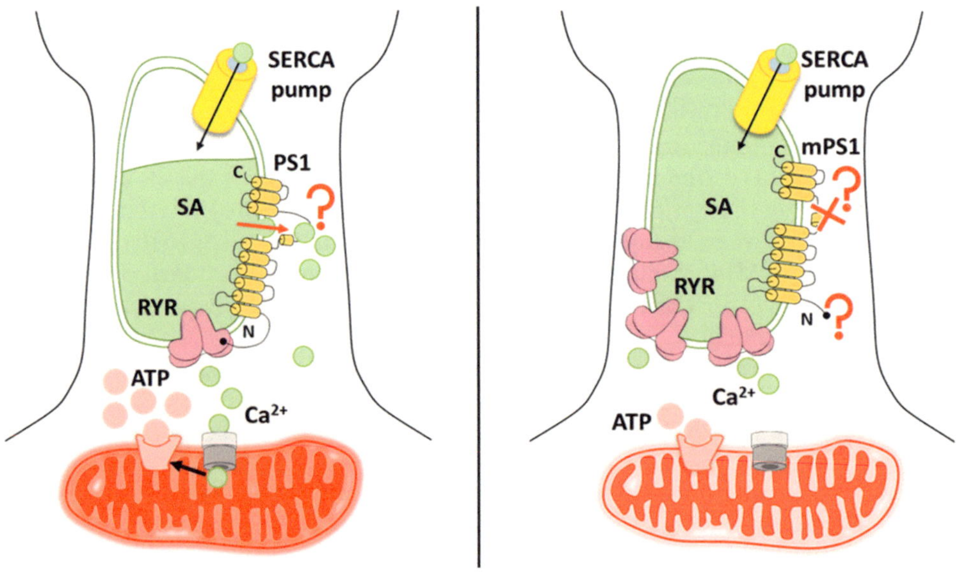


**Figure 3.**
*PS1, possibly, acts as a channel for passive $Ca^{2+}$ leakage from SA and/or may increase $Ca^{2+}$ – induced release of $Ca^{2+}$ by RyR. The N-terminal fragment of PS1 is proposed to bind directly to RyR, increasing its sensitivity to $[Ca^{2+}]i$ and the probability of channel opening. Calcium ions diffuse toward the synaptic mitochondria, entering the matrix through the VDAC/MCU, where they stimulate ATP production. Mutations of PS1 gene possibly abolish the leak function, leading to an overflow of $Ca^{2+}$ reserves in SA. Alternatively, mPS1 may drastically reduce the RyR gating rate and restrict leakage of calcium ions from ER store, through yet unknown mechanisms. Significantly lowered calcium levels released from SA are insufficient for local ATP production by the synaptic mitochondria. The observed increase in RyR expression in mutant PS1 may be a compensatory mechanism to increase calcium leakage from the overfilled store in the absence of correct regulation by presenilin.*

Particularly, disruptions of the mitochondrial energy metabolism are associated with proteins responsible for the production of ATP [81]. In all likelihood, it is in disorders of calcium homeostasis between synaptic mitochondrial clusters and regional calcium stores that one should look for the root causes of many, if not all, neurodegenerative pathologies.

## 10. ER in aging and neurodegeneration

Even in the absence of neurodegeneration, age-related cognitive decline can be observed in association with the loss of synapses in some most active areas of neocortex and hippocampus. Thus, during normal aging, there is a significant decrease in the density of "thin" small-headed spines, which are usually characterized as highly plastic learning spines, in the pyramidal cells of layer III (the most vulnerable cell population in AD) of the prefrontal cortex [82]. The decrease in density, however, has no effect or only minor effect on mushroom spines with normal aging [9]. Whereas in AD, electron microscopy reveals a pathological decrease in the overall density of spines, a decrease in their size along with the presence of abnormally large protrusions in different areas of the brain, as well as morphological abnormalities of the SA and mitochondria (**Figure 4**) [83].

Possibly that a part of the entire pool of thin plastic spines is enriched with ER due to extensive synaptic activation leading to LTP initiation. In its turn, activation could stabilize spine morphology during LTP processes, and increase the head volume, thus, moving them into the pool of mature spines [84, 85]. The remaining poorly activated group of thin spines that did not receive ER, may be completely eliminated. Moreover, the increased and clustered synaptic inputs of such thin highly plastic spines may partially compensate for the bald spots of the synaptic network and improve test performance in plasticity and cognitive abilities [86]. Expectedly, this requires recruiting more mitochondria through local calcium activity. Calcium dysregulation from ER stores may worsen plasticity indicators, reducing the probability of ER penetration into thin, synaptopodin-negative spines, depriving them of the opportunity to achieve energy and mature, while leaving only the perspective of elimination.

In long-term cultured rat hippocampal neurons (15–21 DIV) that mimic physiological aging, a dramatic decrease in SOCE and Orai1/STIM1 is observed, while calcium content in ER stores and the caffeine-induced $Ca^{2+}$ release from the ER are elevated on the contrary. A mechanism of age-dependent SOCE suppression may be proposed to explain the loss of mushroom spines in both physiological aging and cognitive decline in AD [14, 87]. Regulation of the mechanism of SOCE by σ-1R is another important factor in the stabilization of large dendritic spines. Thus, instability of large spines in the hippocampus was observed following dysregulation of the normal SOCE mechanism by σ-1R, in several disorders including AD [88].

Impaired transmission of ER-mitochondrial $Ca^{2+}$ also carries negative physiological consequences for the senescent neurons, since such transmission is necessary to ensure stable ATP flashes from postsynaptic mitochondria upon request. Effective reuptake of $Ca^{2+}$ leakage from ER drops as a result of mitochondrial depolarization in senescent neurons. As a compensatory mechanism for this, cells may develop an increase in the number of MERC contacts, which paradoxically lead to $[Ca^{2+}]$ m overload, disrupting mitochondrial functions and ATP production, respectively [87]. A similar situation is observed in neurodegeneration when the regulation of $Ca^{2+}$ leakage from the ER into mitochondria is disrupted by mutant presenilin,

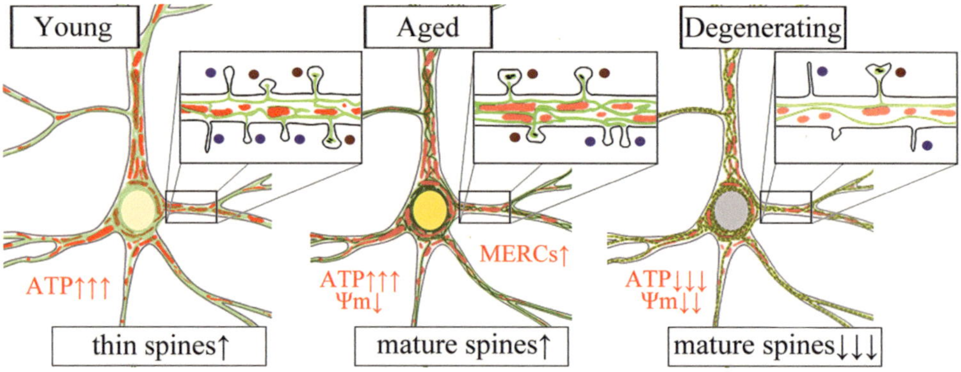


**Figure 4.**
*The ubiquitous mobile ER in young neurons penetrates the pool of thin highly plastic spines more often, thus transforming into spine apparatus and stabilizing the spines. Mitochondria are extensively recruited toward the dendritic region under active postsynaptic compartments to contact ER and make a targeted ATP release. During normal aging, mitochondria drop their resting membrane potential and the ability of addressed ATP production. Consequently, most stable mushroom spines recruit more mitochondria to compensate the lack of ATP/energy by expanding the MERK regions. During neurodegeneration, dysfunctional mitochondria are unable to provide sufficient ATP inflow and calcium sequestration, causing degradation of mature spines and overflow of ER stores with calcium, which leads to a general disruption of ER, malfunction, and, finally, the cell death.*

which pathologically overloads the stores with calcium [89]. However, abnormally increased levels of calcium, when released upon activation, may become inefficient and even toxic for mitochondria that "tuned themselves" to poor consumption of $Ca^{2+}$ in MERC. Such disorganization can be superimposed on the already existing aging-induced functional disorders of MERC/MAM and mitochondrial homeostasis (**Figure 4**).

The multidimensionality of subtle changes in the status of ER stores and their consequences, ER dynamic omnipresence in neurons and close functional relationship with mitochondria as well as other membrane organelles, makes it among the main players in the pathological processes of aging and neurodegeneration.

## 11. Conclusions

1. The spine apparatus is a key structure that is assumed to regulate calcium homeostasis in the postsynaptic region of many synapses. It is characterized by the unique localization of ryanodine receptors, which direct the calcium influx toward the dendritic shaft, the SERCA pumps, and the Orai-STIM complex, which pump calcium ions from the postsynaptic density into the nanoscale store.

2. Synaptic plasticity of many dendritic spines, especially the fraction of the largest ones, depends on stabilization by their spine apparatus. Disturbances in local calcium homeostasis caused by uncoordinated ryanodine/$IP_3$ receptors and machinery for replenishing calcium stores in the ER can lead to the loss of dendritic spines, changes in their morphology, and a decrease in synaptic connectivity.

3. Synaptic mitochondria are characterized by high spatial stability, most often lying directly under the synapses or, more rarely, penetrating into the heads of some large spines. Synapses are the most active consumers of energy, which is

spent on maintaining ion homeostasis, enzymatic processes, and local protein synthesis.

4. Extrasynaptic mitochondria can be transported toward the most active synapses along microtubules, both in anterograde and retrograde directions. It is likely that postsynaptic calcium gradients play the role of spatial markers for "attracting" extrasynaptic mitochondria.

5. Calcium ions also play a fundamental role as signaling agents for the production of ATP by mitochondria. Both intracluster and intercluster competition of mitochondria for synaptic calcium signals is assumed.

6. In all likelihood, calcium gradients arising in the dendritic shaft are pumped into local mitochondria and buffered by them. Rapid and transient calcium gradients occur via release through ryanodine receptors and are regulated by calcium-dependent calcium release mechanisms. These calcium spikelets in synaptic mitochondria can play the role of signals for ATP release, not in a random direction, but toward the source of the gradient.

7. Synaptic clusters, served by the same axonal branch or by "competing axons" from different neuronal sources, are often associated with a complex and underlying mitochondrial cluster. Particular spines in the cluster may compete for locally released ATP. Moreover, mitochondria can efficiently and rapidly buffer released calcium ions, which highly restrict their spread along the dendrite.

8. Presenilin-1 and -2 modulate RyR and regulate the levels of calcium leak from the local storage in the area of mitochondria-associated endoplasmic reticulum membranes (MAM). Mutant presenilins drastically reduce calcium leak at MAM. It leads to overexpression of RYR and overloads the store. The lack of calcium signals from the ER prevents mitochondria from receiving sufficient signals to modulate ATP production and release. As a result, "energy starvation" in the synapses begins, which leads to synaptic deficiency, spine pruning, and neurodegeneration.

## Acknowledgements

The authors thank prof. Menahem Sehgal for constructive comments during the preparation of the manuscript.

## Conflict of interest

The authors declare no conflict of interest.

## Author details

Liliya Kushnireva and Eduard Korkotian*
Department of Brain Sciences, The Weizmann Institute, Rehovot, Israel

*Address all correspondence to: eduard.korkotian@weizmann.ac.il

## References

[1] Verkhratsky A. Physiology and pathophysiology of the calcium store in the endoplasmic reticulum of neurons. Physiological Reviews. 2005;**85**(1):201-279. DOI: 10.1152/physrev.00004.2004

[2] Padamsey Z, Foster WJ, Emptage NJ. Intracellular Ca2+ release and synaptic plasticity: A tale of many stores. The Neuroscientist. 2019;**25**(3):208-226. DOI: 10.1177/1073858418785334

[3] Popov V, Medvedev NI, Davies HA, Stewart MG. Mitochondria form a filamentous reticular network in hippocampal dendrites but are present as discrete bodies in axons: A three-dimensional ultrastructural study. The Journal of Comparative Neurology. 2005;**492**(1):50-65. DOI: 10.1002/cne.20682

[4] Wu Y, Whiteus C, Xu CS, et al. Contacts between the endoplasmic reticulum and other membranes in neurons. Proceedings of the National Academy of Sciences of the United States of America. 2017;**114**(24):E4859-E4867. DOI: 10.1073/pnas.1701078114

[5] Frotscher M, Studer D, Graber W, Chai X, Nestel S, Zhao S. Fine structure of synapses on dendritic spines. Frontiers in Neuroanatomy. 2014;**8**:94. DOI: 10.3389/fnana.2014.00094

[6] Rosado J, Bui VD, Haas CA, Beck J, Queisser G, Vlachos A. Calcium modeling of spine apparatus-containing human dendritic spines demonstrates an "all-or-nothing" communication switch between the spine head and dendrite. PLoS Computational Biology. 2022;**18**(4):e1010069. DOI: 10.1371/journal.pcbi.1010069

[7] Vlachos A. Synaptopodin and the spine apparatus organelle—Regulators of different forms of synaptic plasticity? Annals of Anatomy - Anatomischer Anzeiger. 2012;**194**(4):317-320. DOI: 10.1016/j.aanat.2011.10.014

[8] Ostroff LE, Cain CK, Bedont J, Monfils MH, LeDoux JE. Fear and safety learning differentially affect synapse size and dendritic translation in the lateral amygdala. Proceedings of the National Academy of Sciences of the United States of America. 2010;**107**(20):9418-9423. DOI: 10.1073/pnas.0913384107

[9] Perez-Alvarez A, Yin S, Schulze C, et al. Endoplasmic reticulum visits highly active spines and prevents runaway potentiation of synapses. Nature Communications. 2020;**11**:5083. DOI: 10.1038/s41467-020-18889-5

[10] Sree S, Parkkinen I, Their A, Airavaara M, Jokitalo E. Morphological heterogeneity of the endoplasmic reticulum within neurons and its implications in neurodegeneration. Cell. 2021;**10**(5):970. DOI: 10.3390/cells10050970

[11] Korkotian E, Frotscher M, Segal M. Synaptopodin regulates spine plasticity: Mediation by calcium stores. The Journal of Neuroscience. 2014;**34**(35):11641-11651. DOI: 10.1523/JNEUROSCI.0381-14.2014

[12] Ofer N, Berger DR, Kasthuri N, Lichtman JW, Yuste R. Ultrastructural analysis of dendritic spine necks reveals a continuum of spine morphologies. Developmental Neurobiology. 2021;**81**(5):746-757. DOI: 10.1002/dneu.22829

[13] Deller T, Bas Orth C, Del Turco D, Vlachos A, Burbach GJ, Drakew A, et al. A role for synaptopodin and the spine

apparatus in hippocampal synaptic plasticity. Annals of Anatomy. 2007;**189**(1):5-16. DOI: 10.1016/j.aanat.2006.06.013

[14] Segal M, Korkotian E. Roles of calcium stores and store-operated channels in plasticity of dendritic spines. The Neuroscientist. 2016;**22**(5):477-485. DOI: 10.1177/1073858415613277

[15] Jedlicka P, Deller T. Understanding the role of synaptopodin and the spine apparatus in Hebbian synaptic plasticity - new perspectives and the need for computational modeling. Neurobiology of Learning and Memory. 2017;**138**:21-30. DOI: 10.1016/j.nlm.2016.07.023

[16] Maggio N, Vlachos A. Synaptic plasticity at the interface of health and disease: New insights on the role of endoplasmic reticulum intracellular calcium stores. Neuroscience. 2014;**281**:135-146. DOI: 10.1016/j.neuroscience.2014.09.041

[17] Vlachos A, Ikenberg B, Lenz M, et al. Synaptopodin regulates denervation-induced homeostatic synaptic plasticity. Proceedings of the National Academy of Sciences of the United States of America. 2013;**110**(20):8242-8247. DOI: 10.1073/pnas.1213677110

[18] Leung A, Ohadi D, Pekkurnaz G, Rangamani P. Systems modeling predicts that mitochondria ER contact sites regulate the postsynaptic energy landscape. NPJ Systems Biology and Applications. 2021;7(1):26. DOI: 10.1038/s41540-021-00185-7

[19] Vlachos A, Korkotian E, Schonfeld E, Copanaki E, Deller T, Segal M. Synaptopodin regulates plasticity of dendritic spines in hippocampal neurons. The Journal of Neuroscience. 2009;**29**(4):1017-1033. DOI: 10.1523/JNEUROSCI.5528-08.2009

[20] Lee WC, Bonin V, Reed M, et al. Anatomy and function of an excitatory network in the visual cortex. Nature. 2016;**532**:370-374. DOI: 10.1038/nature17192

[21] Lee KF, Soares C, Thivierge JP, Béïque JC. Correlated synaptic inputs drive dendritic calcium amplification and cooperative plasticity during clustered synapse development. Neuron. 2016;**89**(4):784-799. DOI: 10.1016/j.neuron.2016.01.012

[22] Basnayake K, Mazaud D, Bemelmans A, Rouach N, Korkotian E, Holcman D. Fast calcium transients in dendritic spines driven by extreme statistics. PLoS Biology. 2019;**17**(6):e2006202. DOI: 10.1371/journal.pbio.2006202

[23] Grigoryan G, Segal M. Ryanodine-mediated conversion of STP to LTP is lacking in synaptopodin-deficient mice. Brain Structure & Function. 2016;**221**(4):2393-2397. DOI: 10.1007/s00429-015-1026-7

[24] Segal M. Dendritic spines: Morphological building blocks of memory. Neurobiology of Learning and Memory. 2017;**138**:3-9. DOI: 10.1016/j.nlm.2016.06.007

[25] Korkotian E, Segal M. Synaptopodin regulates release of calcium from stores in dendritic spines of cultured hippocampal neurons. The Journal of Physiology. 2011;**589**(Pt 24):5987-5995. DOI: 10.1113/jphysiol.2011.217315

[26] Kwon HB, Sabatini BL. Glutamate induces de novo growth of functional spines in developing cortex. Nature. 2011;**474**(7349):100-104. DOI: 10.1038/nature09986

[27] Okuno H, Akashi K, Ishii Y, et al. Inverse synaptic tagging of inactive

synapses via dynamic interaction of arc/Arg3.1 with CaMKIIβ. Cell. 2012;**149**(4):886-898. DOI: 10.1016/j.cell.2012.02.062

[28] Kedrov AV, Durymanov M, Anokhin KV. The arc gene: Retroviral heritage in cognitive functions. Neuroscience and Biobehavioral Reviews. 2019;**99**:275-281. DOI: 10.1016/j.neubiorev.2019.02.006

[29] Aloni E, Verbitsky S, Kushnireva L, Korkotian E, Segal M. Increased excitability of hippocampal neurons in mature synaptopodin-knockout mice. Brain Structure and Function. 2021;**226**(7):2459-2466. DOI: 10.1007/s00429-021-02346-0

[30] Yousuf MS, Maguire AD, Simmen T, Kerr BJ. Endoplasmic reticulum-mitochondria interplay in chronic pain: The calcium connection. Molecular Pain. 2020;**16**:1744806920946889. DOI: 10.1177/1744806920946889

[31] Kushnireva L, Korkotian E, Segal M. Calcium sensors STIM1 and STIM2 regulate different calcium functions in cultured hippocampal neurons. Frontiers in Synaptic Neuroscience. 2021;**12**:573714. DOI: 10.3389/fnsyn.2020.573714

[32] Chen YF, Chen LH, Shen MR. The distinct role of STIM1 and STIM2 in the regulation of store-operated Ca2+ entry and cellular function. Journal of Cellular Physiology. 2019;**234**(6):8727-8739. DOI: 10.1002/jcp.27532

[33] Majewski L, Maciąg F, Boguszewski PM, Kuznicki J. Transgenic mice overexpressing human STIM2 and ORAI1 in neurons exhibit changes in behavior and calcium homeostasis but show No signs of neurodegeneration. International Journal of Molecular Sciences. 2020;**21**(3):842. DOI: 10.3390/ijms21030842

[34] Nishiyama J, Yasuda R. Biochemical computation for spine structural plasticity. Neuron. 2015;**87**(1):63-75. DOI: 10.1016/j.neuron.2015.05.043

[35] Lu J, Zuo Y. Clustered structural and functional plasticity of dendritic spines. Brain Research Bulletin. 2017;**129**:18-22. DOI: 10.1016/j.brainresbull.2016.09.008

[36] Redondo R, Morris R. Making memories last: The synaptic tagging and capture hypothesis. Nature Reviews. Neuroscience. 2011;**12**:17-30. DOI: 10.1038/nrn2963

[37] Ostroff LE, Botsford B, Gindina S, et al. Accumulation of polyribosomes in dendritic spine heads, but not bases and necks, during memory consolidation depends on cap-dependent translation initiation. The Journal of Neuroscience. 2017;**37**(7):1862-1872. DOI: 10.1523/JNEUROSCI.3301-16.2017

[38] Rangaraju V, Lauterbach M, Schuman EM. Spatially stable mitochondrial compartments fuel local translation during plasticity. Cell. 2019;**176**(1-2):73-84.e15. DOI: 10.1016/j.cell.2018.12.013

[39] Spillane M, Ketschek A, Merianda TT, Twiss JL, Gallo G. Mitochondria coordinate sites of axon branching through localized intra-axonal protein synthesis. Cell Reports. 2013;**5**(6):1564-1575. DOI: 10.1016/j.celrep.2013.11.022

[40] Rajgor D, Welle TM, Smith KR. The coordination of local translation, membranous organelle trafficking, and synaptic plasticity in neurons. Frontiers in Cell and Development Biology. 2021;**9**:711446. DOI: 10.3389/fcell.2021.711446

[41] Godoy JA, Arrázola MS, Ordenes D, Silva-Alvarez C, Braidy N, Inestrosa NC. Wnt-5a ligand modulates mitochondrial

fission-fusion in rat hippocampal neurons. The Journal of Biological Chemistry. 2014;**289**(52):36179-36193. DOI: 10.1074/jbc.M114.557009

[42] Faitg J, Lacefield C, Davey T, White K, Laws R, Kosmidis S, et al. 3D neuronal mitochondrial morphology in axons, dendrites, and somata of the aging mouse hippocampus. Cell Reports. 2021;**36**(6):109509. DOI: 10.1016/j.celrep.2021.109509

[43] Lewis TL, Kwon SK, Lee A, et al. MFF-dependent mitochondrial fission regulates presynaptic release and axon branching by limiting axonal mitochondria size. Nature Communications. 2018;**9**:5008. DOI: 10.1038/s41467-018-07416-2

[44] Freeman DW, Petralia RS, Wang YX, Mattson MP, Yao PJ. Mitochondria in hippocampal presynaptic and postsynaptic compartments differ in size as well as intensity. Matters (Zur). 2017;**2017**. DOI: 10.19185/matters.201711000009

[45] Chang DT, Honick AS, Reynolds IJ. Mitochondrial trafficking to synapses in cultured primary cortical neurons. The Journal of Neuroscience. 2006;**26**(26):7035-7045. DOI: 10.1523/JNEUROSCI.1012-06.2006

[46] Li Z, Okamoto K, Hayashi Y, Sheng M. The importance of dendritic mitochondria in the morphogenesis and plasticity of spines and synapses. Cell. 2004;**119**(6):873-887. DOI: 10.1016/j.cell.2004.11.003

[47] Eisner V, Csordás G, Hajnóczky G. Interactions between sarco-endoplasmic reticulum and mitochondria in cardiac and skeletal muscle - pivotal roles in $Ca^{2+}$ and reactive oxygen species signaling. Journal of Cell Science. 2013;**126**(Pt 14):2965-2978. DOI: 10.1242/jcs.093609

[48] Shim SH, Xia C, Zhong G, et al. Super-resolution fluorescence imaging of organelles in live cells with photoswitchable membrane probes. Proceedings of the National Academy of Sciences of the United States of America. 2012;**109**(35):13978-13983. DOI: 10.1073/pnas.1201882109

[49] Liu X, Weaver D, Shirihai O, Hajnóczky G. Mitochondrial 'kiss-and-run': Interplay between mitochondrial motility and fusion-fission dynamics. The EMBO Journal. 2009;**28**(20):3074-3089. DOI: 10.1038/emboj.2009.255

[50] Sheng ZH, Cai Q. Mitochondrial transport in neurons: Impact on synaptic homeostasis and neurodegeneration. Nature Reviews. Neuroscience. 2012;**13**:77-93. DOI: 10.1038/nrn3156

[51] Wang X, Schwarz TL. The mechanism of Ca2+ -dependent regulation of kinesin-mediated mitochondrial motility. Cell. 2009;**136**(1):163-174. DOI: 10.1016/j.cell.2008.11.046

[52] Wang X, Winter D, Ashrafi G, et al. PINK1 and Parkin target Miro for phosphorylation and degradation to arrest mitochondrial motility. Cell. 2011;**147**(4):893-906. DOI: 10.1016/j.cell.2011.10.018

[53] Macaskill AF, Rinholm JE, Twelvetrees AE, et al. Miro1 is a calcium sensor for glutamate receptor-dependent localization of mitochondria at synapses. Neuron. 2009;**61**(4):541-555. DOI: 10.1016/j.neuron.2009.01.030

[54] Gerencser AA, Chinopoulos C, Birket MJ, et al. Quantitative measurement of mitochondrial membrane potential in cultured cells: Calcium-induced de- and hyperpolarization of neuronal mitochondria. The Journal of Physiology. 2012;**590**(12):2845-2871. DOI: 10.1113/jphysiol.2012.228387

[55] Vos M, Lauwers E, Verstreken P. Synaptic mitochondria in synaptic transmission and organization of vesicle pools in health and disease. Frontiers in Synaptic Neuroscience. 2010;**2**:139. DOI: 10.3389/fnsyn.2010.00139

[56] Chamberland S, Moratallaab AZ, Topolnik L. Calcium extrusion mechanisms in dendrites of mouse hippocampal CA1 inhibitory interneurons. Cell Calcium. 2019;**77**:49-57. DOI: 10.1016/j.ceca.2018.12.002

[57] Vianello A, Casolo V, Petrussa E, et al. The mitochondrial permeability transition pore (PTP) — An example of multiple molecular exaptation? Biochimica et Biophysica Acta (BBA) – Bioenergetics. 2012;**1817**(11):2072-2086. DOI: 10.1016/j.bbabio.2012.06.620

[58] Duchen MR, Leyssens A, Crompton M. Transient mitochondrial depolarizations reflect focal sarcoplasmic reticular calcium release in single rat cardiomyocytes. The Journal of Cell Biology. 1998;**142**(4):975-988. DOI: 10.1083/jcb.142.4.975

[59] Levy M, Faas GC, Saggau P, Craigen WJ, Sweatt JD. Mitochondrial regulation of synaptic plasticity in the hippocampus. The Journal of Biological Chemistry. 2003;**278**(20):17727-17734. DOI: 10.1074/jbc.M212878200

[60] Chang KT, Niescier RF, Min KT. Mitochondrial matrix Ca2+ as an intrinsic signal regulating mitochondrial motility in axons. Proceedings of the National Academy of Sciences of the United States of America. 2011;**108**(37):15456-15461. DOI: 10.1073/pnas.1106862108

[61] Giacomello M, Pellegrini L. The coming of age of the mitochondria-ER contact: A matter of thickness. Cell Death and Differentiation. 2016;**23**(9):1417-1427. DOI: 10.1038/cdd.2016.52

[62] Rizzuto R, De Stefani D, Raffaello A, Mammucari C. Mitochondria as sensors and regulators of calcium signalling. Nature Reviews. Molecular Cell Biology. 2012;**13**(9):566-578. DOI: 10.1038/nrm3412

[63] Weng TY, Tsai SYA, Su TP. Roles of sigma-1 receptors on mitochondrial functions relevant to neurodegenerative diseases. Journal of Biomedical Science. 2017;**24**:74. DOI: 10.1186/s12929-017-0380-6

[64] Ryskamp DA, Zhemkov V, Bezprozvanny I. Mutational analysis of Sigma-1 Receptor's role in synaptic stability. Frontiers in Neuroscience. 2019;**13**:1012. DOI: 10.3389/fnins.2019.01012

[65] Zhemkov V, Geva M, Hayden MR, Bezprozvanny I. Sigma-1 receptor (S1R) interaction with cholesterol: Mechanisms of S1R activation and its role in neurodegenerative diseases. International Journal of Molecular Sciences. 2021;**22**(8):4082. DOI: 10.3390/ijms22084082

[66] Jia H, Zhang Y, Huang Y. Imaging sigma receptors in the brain: New opportunities for diagnosis of Alzheimer's disease and therapeutic development. Neuroscience Letters. 2019;**691**:3-10. DOI: 10.1016/j.neulet.2018.07.033

[67] Dromard Y, Arango-Lievano M, Fontanaud P, Tricaud N, Jeanneteau F. Dual imaging of dendritic spines and mitochondria in vivo reveals hotspots of plasticity and metabolic adaptation to stress. Neurobiology Stress. 2021;**15**:100402. DOI: 10.1016/j.ynstr.2021.100402

[68] Perkins GA, Renken CW, Frey TG, Ellisman MH. Membrane architecture of mitochondria in neurons of the central

nervous system. Journal of Neuroscience Research. 2001;**66**(5):857-865. DOI: 10.1002/jnr.10050

[69] Stutzmann GE, Smith I, Caccamo A, Oddo S, Laferla FM, Parker I. Enhanced ryanodine receptor recruitment contributes to Ca2+ disruptions in young, adult, and aged Alzheimer's disease mice. The Journal of Neuroscience. 2006;**26**(19):5180-5189. DOI: 10.1523/JNEUROSCI.0739-06.2006

[70] Tu H, Nelson O, Bezprozvanny A, et al. Presenilins form ER Ca2+ leak channels, a function disrupted by familial Alzheimer's disease-linked mutations. Cell. 2006;**126**(5):981-993. DOI: 10.1016/j.cell.2006.06.059

[71] Del Prete D, Checler F, Chami M. Ryanodine receptors: Physiological function and deregulation in Alzheimer disease. Molecular Neurodegeneration. 2014;**9**:21. DOI: 10.1186/1750-1326-9-21

[72] Lee SY, Hwang DY, Kim YK, et al. PS2 mutation increases neuronal cell vulnerability to neurotoxicants through activation of caspase-3 by enhancing of ryanodine receptor-mediated calcium release. The FASEB Journal. 2006;**20**(1):151-153. DOI: 10.1096/fj.05-4017fje;1

[73] Rybalchenko V, Hwang SY, Rybalchenko N, Koulen P. The cytosolic N-terminus of presenilin-1 potentiates mouse ryanodine receptor single channel activity. The International Journal of Biochemistry & Cell Biology. 2008;**40**(1):84-97. DOI: 10.1016/j.biocel.2007.06.023

[74] Gao P, Yan Z, Zhu Z. Mitochondria-associated endoplasmic reticulum membranes in cardiovascular diseases. Frontiers in Cell and Development Biology. 2020;**8**:604240. DOI: 10.3389/fcell.2020.604240

[75] De la Fuente S, Sheu SS. SR-mitochondria communication in adult cardiomyocytes: A close relationship where the Ca2+ has a lot to say. Archives of Biochemistry and Biophysics. 2019;**663**:259-268. DOI: 10.1016/j.abb.2019.01.026

[76] Chami M, Checler F. Alterations of the endoplasmic reticulum (ER) calcium signaling molecular components in Alzheimer's disease. Cell. 2020;**9**(12):2577. DOI: 10.3390/cells9122577

[77] Gazda K, Kuznicki J, Wegierski T. Knockdown of amyloid precursor protein increases calcium levels in the endoplasmic reticulum. Scientific Reports. 2017;**7**(1):14512. DOI: 10.1038/s41598-017-15166-2

[78] Du H, Guo L, Yan S, Sosunov AA, McKhann GM, Yan SS. Early deficits in synaptic mitochondria in an Alzheimer's disease mouse model. Proceedings of the National Academy of Sciences of the United States of America. 2010;**107**(43):18670-18675. DOI: 10.1073/pnas.1006586107

[79] Müller WE, Eckert A, Kurz C, et al. Mitochondrial dysfunction: Common final pathway in brain aging and Alzheimer's disease—Therapeutic aspects. Molecular Neurobiology. 2010;**41**:159-171. DOI: 10.1007/s12035-010-8141-5

[80] Smit-Rigter L, Rajendran R, Silva CA, et al. Mitochondrial dynamics in visual cortex are limited In vivo and not affected by axonal structural plasticity. Current Biology. 2016;**26**(19):2609-2616. DOI: 10.1016/j.cub.2016.07.033

[81] Yu H, Lin X, Wang D, et al. Mitochondrial molecular abnormalities revealed by proteomic analysis of

hippocampal organelles of mice triple transgenic for Alzheimer disease. Frontiers in Molecular Neuroscience. 2018;**11**:74. DOI: 10.3389/fnmol.2018.00074

[82] Dumitriu D, Hao J, Hara Y, et al. Selective changes in thin spine density and morphology in monkey prefrontal cortex correlate with aging-related cognitive impairment. The Journal of Neuroscience. 2010;**30**(22):7507-7515. DOI: 10.1523/JNEUROSCI.6410-09.2010

[83] Baloyannis SJ. What has electron microscopy contributed to Alzheimer's research? Future Neurology. 2015;**10**(6): 515-527. DOI: 10.2217/fnl.15.44

[84] Korkotian E, Segal M. Structure-function relations in dendritic spines: Is size important? Hippocampus. 2000;**10**(5):587-595. DOI: 10.1002/1098-1063(2000)10:5<587::AID-HIPO9>3.0.CO;2-C

[85] Basnayake K, Mazaud D, Kushnireva L, Bemelmans A, Rouach N, Korkotian E, et al. Nanoscale molecular architecture controls calcium diffusion and ER replenishment in dendritic spines. Science Advances. 2021;**7**(38):eabh1376. DOI: 10.1126/sciadv.abh1376

[86] Pereira AC, Lambert HK, Grossman YS, et al. Glutamatergic regulation prevents hippocampal-dependent age-related cognitive decline through dendritic spine clustering. Proceedings of the National Academy of Sciences of the United States of America. 2014;**111**(52):18733-18738. DOI: 10.1073/pnas.1421285111

[87] Calvo-Rodríguez M, García-Durillo M, Villalobos C, Núñez L. In vitro aging promotes endoplasmic reticulum (ER)-mitochondria Ca2+ cross talk and loss of store-operated Ca2+ entry (SOCE) in rat hippocampal neurons. Biochimica et Biophysica Acta. 2016;**1863**(11):2637-2649. DOI: 10.1016/j.bbamcr.2016.08.001

[88] Ryskamp DA, Korban S, Zhemkov V, Kraskovskaya N, Bezprozvanny I. Neuronal Sigma-1 receptors: Signaling functions and protective roles in neurodegenerative diseases. Frontiers in Neuroscience. 2019;**13**:862. DOI: 10.3389/fnins.2019.00862

[89] Sun S, Zhang H, Liu J, et al. Reduced synaptic STIM2 expression and impaired store-operated calcium entry cause destabilization of mature spines in mutant presenilin mice. Neuron. 2014;**82**(1):79-93. DOI: 10.1016/j.neuron.2014.02.019

Chapter 5

# The Role of Endoplasmic Reticulum Stress and Its Regulation in the Progression of Neurological and Infectious Diseases

*Mary Dover, Michael Kishek, Miranda Eddins, Naneeta Desar, Ketema Paul and Milan Fiala*


## Abstract

The unfolded protein response (UPR) is a cellular mechanism activated by endoplasmic reticulum (ER) stress, which ranges from inhibition of protein synthesis to apoptosis. ER stress is induced in general by aggregated autologous or foreign (e.g. viral) proteins, oxidative stress, mitochondrial dysfunction, disruption of intracellular calcium, or inflammation. In patients with Alzheimer's disease (AD) and amyotrophic lateral sclerosis (ALS), the known stressors are aggregated amyloid-beta and superoxide dismutase (SOD-1), respectively, but autologous DNA released by trauma into the cytoplasm may also be involved in ALS. In HIV-1-associated neurocognitive disorders (HAND), ER stress is induced by HIV-1 and antiretroviral therapy. Additionally, in cases of epilepsy, ER stress has been implicated in neuronal dysfunction. In this chapter, we examine a clinical and immunologic approach to ER stress in the progression of neurological and infectious diseases. In addition, we will briefly discuss emerging treatments including omega fatty acids, progesterone, and DHA, which repair and favorably regulate UPR in some patients with neurological diseases.


**Keywords:** ER stress, UPR, Alzheimer's disease, amyotrophic lateral sclerosis, HIV, epilepsy

## 1. Introduction

For decades, researchers in a diverse array of scientific fields have been working to understand the pathogenesis of neurodegenerative diseases including Alzheimer's disease (AD) and amyotrophic lateral sclerosis (ALS). However, there is still much discourse within the scientific community on what exactly causes such diseases to occur. In recent years, the endoplasmic reticulum (ER), has been implicated in the progression of neurological diseases [1] including AD, ALS, epilepsy, and human immunodeficiency virus (HIV)-associated neurocognitive disorder (HAND). To understand how exactly this cellular organelle can lead to such detrimental neurological disorders, a basic understanding of the function of the ER is necessary.

The ER is a cellular organelle that functions to fold proteins correctly following their translation by ribosomes. The ER also plays an integral role in physiological homeostasis, primarily via calcium regulation [2] and protein synthesis [3]. Typically, correctly folded proteins are secreted by the ER and are transported to the Golgi body for further processing and sorting for transport to their eventual destinations. However, ER overload results in protein misfolding and accumulation of unfolded proteins in the ER lumen, which induces ER stress [4]. ER stress signals the cell to enter the survival pathway and initiates an unfolded protein response (UPR) [5] through three ER membrane-associated proteins: inositol requiring enzyme 1 (IRE1), pancreatic ER kinase (PERK), and activating transcription factor-6 (ATF6) [6].

IRE1 (encoded by the *ERN1* gene) is an ER transmembrane receptor sensitive to misfolded protein aggregation in the ER lumen [7]. Activated IRE1 stimulates the translation of the X-box binding protein 1 (XBP1) into a transcription factor, which upregulates ER chaperones and ER-associated degradation (ERAD) [8]. Ideally in a functioning ER, these ERAD elements will assist with the degradation of unfolded proteins within the ER lumen to reduce ER stress.

PERK reduces protein synthesis by inactivation of eukaryotic initiation factor 2α, leading either to a pro-apoptotic pathway or a protective pathway involving chaperones and foldases [9]. Thus, PERK expression will mitigate the additional buildup of proteins within the ER. However, in extreme conditions, the cell suffers apoptosis in response to excessive misfolded protein aggregation [10].

Finally, ATF6 increases the transcription of several ER proteins including chaperones, foldases, and ERAD elements. These ER proteins increase the cell's ability to fold proteins and reduce the load of unfolded proteins in the ER [11].

These ER stress proteins communicate between ER and mitochondria in the region called the mitochondria-associated ER membrane [9]. Activation of all three of these transmembrane proteins is prompted by unfolded protein buildup in the lumen of the ER, which in turn activates the nuclear transcription of ERAD elements (**Figure 1**).

Thus, the role of ER stress in the progression of AD, ALS, epilepsy, and HAND are of rising interest in the scientific community and may serve as a possible explanation

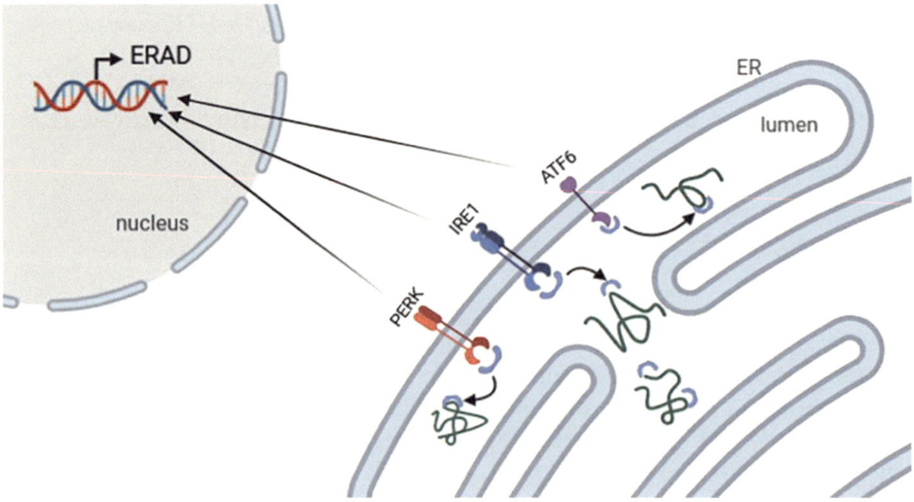


**Figure 1.**
*A scheme of the unfolded protein response in cells with ER stress. Illustration created with Biorender.com.*

for their development. Due to the increasing evidence implicating ER stress in these diseases, researchers have shifted their focus to studying possible treatments that target the ER to mitigate disease progression. In this chapter, we will discuss the role of ER stress in AD, ALS, epilepsy, and HAND, and review the current therapeutic options for treating ER stress such as omega fatty acids including docosahexaenoic acid (DHA), and progesterone.

## 2. ER stress and the UPR in Alzheimer's disease

AD is a progressive neurodegenerative disease characterized by the buildup of the cytotoxic protein aggregates amyloid-beta (Aβ) and fibrillary phospho-tau (P-τ). Within the brain, Aβ accumulates into plaques that can block synaptic function [12], induce ER stress [13], and ultimately cause neuronal apoptosis [14]. These cytotoxic proteins ultimately lead to widespread neurodegeneration leading to cognitive decline [12]. AD can arise via genetic factors in familial AD (fAD) or lifestyle factors in sporadic AD (sAD), but ER stress has been shown to play a role in the progression of both cases.

Many factors including oxidative stress, mitochondrial dysfunction, glutamate-induced neurotoxicity, and imbalance of calcium contribute to ER stress and UPR induction in AD patients [15].

Oxidative stress in the cell often results from the buildup of reactive oxygen species (ROS) from dysfunctional mitochondria [16] and can induce ER stress. In patients with AD, an imbalance in antioxidants can lead to an increase in ROS which can cause widespread cell damage [16]. In normal aging, this antioxidant imbalance is increasingly common and the body is often pushed into a "pro-oxidative state" meaning that the level of ROS is higher than normal [17]. However, in patients with AD, the aggregation of Aβ exacerbates this pro-oxidative state [18]. A 2001 study revealed that Aβ's interaction with Iron within the brain causes increased ROS, and by treating aggregated Aβ with an Iron chelator there is a marked reduction of neural toxicity [18]. Typically, ROS levels are mediated by the ER proteins PERK and ATF6. However, when ROS levels exceed the normal range, PERK and ATF6 are unable to produce enough antioxidants, and the ER enters a state of ER stress [17], exacerbating cellular dysfunction and leading to neuronal apoptosis (**Figure 2**).

Additionally, glutamate has been implicated in the buildup of ROS in the AD brain. When glutamate binds to NMDA receptors, this prompts glutamate-induced neurotoxicity (GNT). GNT causes a major influx of calcium into the cell and prompts the release of ROS from the mitochondria [19]. A buildup of GNT in the AD brain can impair glutamate receptors necessary for metabolism [20], lead to increased production of ROS, and cause an imbalance of cytosolic calcium leading to cell death (**Figure 2**) [19].

However, the exact etiology of AD concerning ER stress is still under investigation. Recent studies have pointed to the buildup of Aβ and P-τ as the cause of the development of ER stress and the UPR [13]. On the other hand, some believe that ER stress arises from increased ROS in the brain, leading to the buildup of Aβ and P-τ, which are released into the brain following cellular apoptosis [21]. Therefore, the sequence of biochemical events in AD is still very much under debate.

Additionally, the presenilin proteins have been implicated in ER-stress-mediated AD pathogenesis. Presenilin 1 (PS1) and presenilin 2 (PS2) are part of the γ-secretase complex which mediates the cleavage of the amyloid-beta precursor protein (APP) [22]. In fAD, mutations in these proteins have been found to alter the amyloid

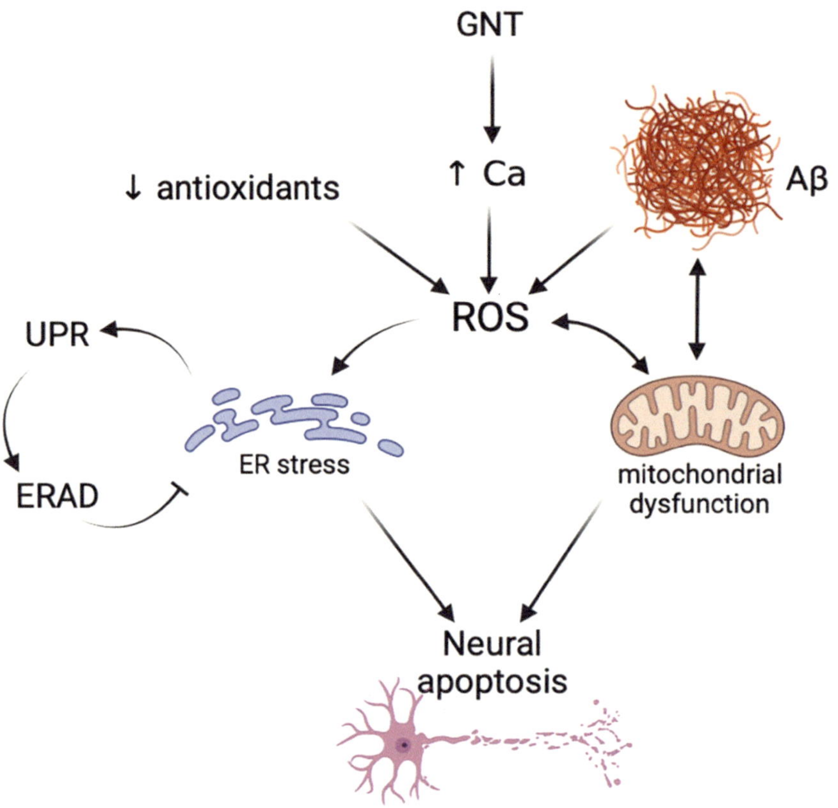


**Figure 2.**
*A scheme depicting the effect of decreased antioxidants, glutamate neurotoxicity (GNT), and amyloid-beta (Aβ) aggregation on the buildup of ROS and ER stress in AD patients. Image created with Biorender.com.*

precursor protein (APP) cleavage process, resulting in higher levels of cytotoxic Aβ [21]. The mutant PS1 affects the ER stress response attributed to the inhibited activation of the ER stress pathways IRE1, PERK, and ATF6. Cells expressing PS1 mutants also display increased Aβ production and increased sensitivity to apoptosis caused by ER stress [21]. Therefore, the damage associated with the PS1 mutant proteins cannot be reversed by chaperones and folding proteins, and apoptosis is the most common outcome. In patients with sAD, mutations in the PS2 gene are also linked to disease progression by downregulation of the UPR pathways [23].

The target genes of the transcription factor XBP1 are also linked to AD. XBP1 affects the expression of at least one of the key proteins in the γ-secretase complex, primarily UBQLN1, the gene coding for ubiquitin, which is a negative regulator of presenilins. UBQLN1 plays a role in the control of APP trafficking. Therefore, in the production of Aβ, reduced expression of XBP1 in AD increases the production of Aβ and causes apoptosis [24].

Thus, the combined effects of a buildup of ROS, mitochondrial dysfunction, GNT, and presenilin mutation exacerbate the effects of normal aging in patients with AD and lead to an increase in ER stress. Thus, ER stress has been shown to play an integral role in the pathogenesis of AD through several pathways.

## 3. ER stress in epilepsy

Epilepsy is a neurological disease that involves chronic seizures as well as abnormal brain activity, which causes periods of unusual behavior or sensations [25]. There are different types of epileptic seizures, including generalized, focal, and unknown onset epilepsy, which differ in the area in which seizures occur [26]. The most common type of epilepsy among adults is focal epilepsy known as temporal lobe epilepsy (TLE), however, patients who continue to experience seizures are at risk of other areas of the brain becoming damaged, such as the hippocampus [27].

Current research incriminates ER stress as an important factor in the pathogenesis of epilepsy [25]. Neuronal death after a seizure is caused by apoptotic signaling pathways, of which there have been two gene families identified to be critical. These gene families are the Bcl-2 family proteins and the caspases [27], which have both been linked to ER stress generation [28]. Bcl-2 proteins are apoptosis regulators, and caspases are proteases that cleave certain proteins and enzymes that induce apoptosis. A study on the hippocampus of patients with epilepsy in 2006 revealed that caspases 6, 7, and 9 were higher in patients with epilepsy than in controls and were localized to the ER-containing region of the temporal lobe [27]. Additionally, patients' hippocampal regions contained altered levels of Bcl-2 protein and an increase in both the ER stress-related motif KDEL and calnexin [27]. KDEL and calnexin are markers for ER stress and were used here to examine levels of ER stress in epilepsy patients [27, 29]. Finally, the researchers showed co-localization of the ER stress marker KDEL and caspases within the epileptic patients' brain [27]. Thus, this study found that apoptotic pathways in epilepsy are directly linked to the ER and can induce ER stress.

The ER stress-related pathways discussed earlier were also abnormal in animal studies. In a rat model of status epilepticus (SE), phosphorylated PERK (p-PERK), phosphorylated eIF2α (p-eIF2α), and C/EBP homologous protein (*CHOP*) [30] were increased. Although the *CHOP* protein was considered as causing cell death, it has been reinterpreted as protective against neuronal death after a seizure [31]. In another study, the hippocampus of mice with status epilepticus (SE) was found to have elevated mRNA levels of spliced XBP1 compared to controls [31], suggesting an activation of the IRE1 branch. Higher XBP1 levels lead to an increase in the expression of chaperones for ER stress reduction. In the same study, it was found that these mice also had increased levels of ATF in the hippocampus [31]. Therefore, ER stress in epilepsy appears to induce both injurious pro-apoptotic and beneficial anti-apoptotic pathways.

## 4. ER stress in amyotrophic lateral sclerosis

The neurologic disorder ALS is characterized by its neuropathology demonstrating neuronal apoptosis in the spinal cord and brain, and inflammatory attack of neurons induced by aggregated superoxide dismutase (SOD-1) and other stimuli [32, 33]. The progression of ALS is associated with a gradual decrease in movement, paralysis, and overall weakness [34]. ALS has a sporadic form (sALS) with an unknown cause and a familial form (fALS) [35].

ER stress in individuals with fALS can be triggered by the Cu/Zn SOD-1 protein malfunction. This protein is responsible for degrading ROS, therefore, a mutation in SOD-1 promotes the disease [36]. As previously described, the increase of ROS results

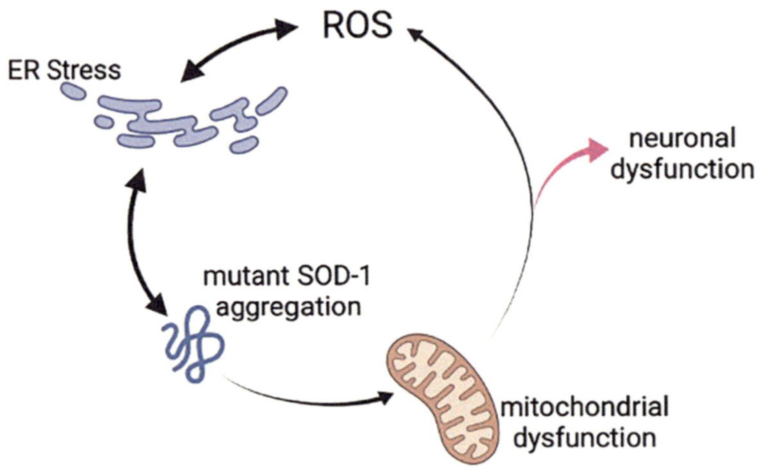


**Figure 3.**
*A scheme depicting the cycle of ER stress buildup in ALS. Illustration created with Biorender.com.*

from the increase in GNT and a greater amount of calcium within the cell, damaging both the ER and mitochondria [1]. Disease progression of ALS is identified by mitochondrial dysfunction and a change in the mitochondrial membrane structure due to a release in molecules that initiate apoptosis [21]. Due to the aggregation of mutated SOD1 in the mitochondria of motor neurons, motor skills are adversely affected [35]. The protein previously implicated in neuronal apoptosis in epilepsy, Bcl-2, is also responsible for maintaining the structural integrity of the mitochondrial membrane [37]. When mutated SOD-1 interacts with the Bcl-2 protein in ALS patients, the membrane is weakened. Thus, there is an increase in ROS production due to the mutation of the SOD-1 protein creating an increase in ER stress in individuals with ALS (**Figure 3**) [21].

In ALS individuals, caspase activation is also common [38]. We know from caspase's role in epilepsy that caspases have been implicated in neuronal apoptosis and ER stress genesis [28]. Caspase activation is especially prevalent when the SOD1 mutation is present. In a mouse model of mutant Cu/Zn SOD-1 exhibiting symptoms of ALS, an increase in oxidative stress was related to activation of caspase 12, 9, and 3 in the spinal cord [38]. Caspase 12 is located in the endoplasmic reticulum [39] wherein the cleavage of this enzyme overall increases the oxidative stress further causing an exaggeration of ALS symptoms in the mice. This cleavage of caspase-12 may be a result of the activation of calpain, a calcium-dependent enzyme, in the mice's spinal cord [21].

Thus, SOD-1 aggregates in individuals with ALS exacerbate ROS release and mitochondrial dysfunction, which induces ER stress and leads to eventual neuronal apoptosis.

## 5. ER stress in human immunodeficiency virus 1

Human immunodeficiency virus 1 (HIV-1) is the result of acquired immunodeficiency syndrome (AIDS), characterized by the destruction of CD4 T cells [40]. In addition to the dramatic opportunistic infection *Pneumocystis carinii* pneumonia, AIDS is associated with degenerative complications in the brain and the heart, HIV-1

*DOI: http://dx.doi.org/10.5772/intechopen.105543*

encephalitis, and cardiomyopathy, which display inflammatory damage by virus-infected monocyte/macrophages [40]. HIV-1 particles are carried into the brain and the heart in monocytes through interendothelial gaps opened by virus envelope gp120, TNF-α, and the antiretroviral (ART) drug azidothymidine [41]. Because of this, besides encephalitis, the neuropathology of HIV-1 includes subtler HIV-associated neurocognitive disorders (HAND) [42].

ER stress and UPR activation play a large role in the neurodegeneration of AIDS and HIV-1 patients and, thus, are a potential therapeutic target. In HAND, the expression of the three UPR pathways in astrocytes leads to the opening of the mitochondrial permeability transition pore (mPTP) linked to apoptosis [42]. The inflammatory cytokine IL-1β along with ART drugs increases cytosolic calcium and triggers ER stress through upregulation of UPR pathways, mitochondrial depolarization, excitotoxicity, and increased ROS [43]. A recent study in 2020 examined the effects of ART on the ER stress pathway in HAND. The researchers found that HIV along with IL-1β increased the UPR transcripts IRE1, PERK, and ATF6. Another HAND signal, the nucleotide reverse transcriptase inhibitor (NRTI) abacavir, upregulated AEG-1 transcription and regulated calcium signaling and ER quality control by co-localizing with calnexin, an integral calcium-dependent chaperon protein in the ER [43]. Together, IL-1B and abacavir induce increased intracellular calcium (via ER calcium release) in astrocytes to levels comparable to that of the known ER stressor thapsigargin [43]. A prolonged intracellular increase in calcium triggers mitochondrial depolarization and ER stress response, as it increases mitochondrial permeability transition pore (mPTP). The mPTP opening leads to increased ROS as well as calcium-dependent exocytosis of glutamate, which causes GNT and neuronal damage experienced in HAND [43].

Additionally, HIV-1 upregulates the Tat protein, which induces GNT, ER stress, mitochondrial dysfunction, and UPR [44]. Therefore, astrocyte ER stress could act as a therapeutic target for HIV-1 neuronal infection: IL-1β and abacavir induce intracellular calcium dysregulation and mitochondrial dysfunction (mPTP opening), which can lead to apoptosis via triggering ER stress and increase in UPR signaling pathways [43].

## 6. Therapies for treating ER stress in neurodegenerative patients

Due to its implications in a variety of neurodegenerative diseases, ER stress and UPR proteins are growing targets for immunotherapy treatment for several neurological disorders. Various lipid-based molecules such as omega-3 fatty acids and progesterone induce neuroprotective effects against ER stress, regulating Aβ-induced neuroinflammation.

Omega-3 fatty acids have been found to modulate UPR counteracting ER stress. In macrophages of patients with AD and mild cognitive impairment (MCI), *in vitro* supplementation with fish-derived ω-3 fatty acids resulted in downregulation of ER stress signature genes *CHOP, DDIT3*, and *CASP3* (caspase-3), which typically promote apoptosis [15]. Further, fatty acid treatment upregulated genes associated with UPR proteins including *IRE1*, *ATF6*, and *ATF4* [15]. Thus, the researchers concluded that the phospho-PERK pathway was heterogeneously affected by omega-3 fatty acids via an increase in immunoenhancement but a decrease in pro-apoptosis, but omega-3 can enhance UPR genes necessary for combating ER stress in patients with AD [15].

Docosahexaenoic acid (DHA), a specific omega-3 fatty acid beneficially responds to the ER stress induced by traumatic brain injury (TBI), which triggers calcium homeostatic disruption. The mechanism by which DHA prevents ER stress is through its protectins, which block ER stress-inducing $IP_3R$-mediated ER $Ca^{2+}$ depletion [45]. Protectins such as neuroprotection D1 (NPD1), a DHA derivative, upregulate anti-apoptotic factors and downregulate pro-apoptotic factors [45]. In addition, DHA's resolving derivatives also directly decrease the inflammation responses to ER stress by blocking pro-inflammatory cytokines TNF-α and ILβ1 as well as reducing pro-inflammatory mediators prostaglandin E2, thromboxanes, and leukotrienes [45]. DHA administration reduced post-TBI increase of CHOP-gene expressing microglia and macrophages [46]. Thereby, DHA as an omega-3 fatty is seen to decrease the triggering and the effects of ER stress through various mechanisms and is, therefore, a potential therapy.

The neurosteroid progesterone has additionally been seen to improve ER stress in its applications in AD astrocytes with the release of IL-1 and TNF-α [47]. In these AD cells, Aβ-induced ER stress and inflammation were mediated by progesterone. Progesterone is associated with the downregulation of pro-inflammatory cytokines which in turn reduces ER stress activation. This effect is further shown through progesterone's attenuation of GRP78 expression, which is implicated in amyloid-beta-induced ER stress response [47].

Though therapies targeting neurodegenerative diseases such as ALS or epilepsy have not implicated ER stress as a potential therapeutic target, substantial research on ER stress treatment in AD brains has uncovered omega-3 fatty acids such as DHA and neurosteroid progesterone as potential supplemental therapies for ER stress and neuroinflammation.

## 7. Conclusions

ER stress has been implicated in several neurological and infectious diseases, primarily AD, ALS, epilepsy, and HIV-1. In AD, mutant PS1 and PS2 induced amyloid-beta aggregation and an increase in ROS lead to ER stress and neuronal death. In ALS, aggregated SOD-1 increases the buildup of ROS in the cytoplasm, inducing ER stress. Patients with epilepsy have also shown elevated ER stress with increased levels of pro-apoptotic chaperones, yet the exact mechanism is still under investigation. Finally, HIV-1-associated neurocognitive decline leads to a buildup of the pro-inflammatory cytokine IL-1B which induces ER stress and exacerbates neurocognitive decline. The apparent role of ER stress in the progression of these neurological and infectious diseases has prompted scientists to explore treatments targeting ER stress and the UPR. Treatment with omega fatty acids, progesterone, and DHA have shown positive effects on ER stress levels in patients with AD, yet a true treatment has not been developed.

## Conflict of interest

The authors declare no conflict of interest.

## Author details

Mary Dover, Michael Kishek, Miranda Eddins, Naneeta Desar, Ketema Paul
and Milan Fiala*
Department of Integrative Physiology and Biology, UCLA School of Life Sciences,
Los Angeles, California, USA

*Address all correspondence to: mfiala@ucla.edu

## References

[1] Lindholm D, Wootz H, Korhonen L. ER stress and neurodegenerative diseases. Cell Death and Differentiation. 2006;**13**(3):385-392

[2] Rasheva VI, Domingos PM. Cellular responses to endoplasmic reticulum stress and apoptosis. Apoptosis. 2009;**14**(8):996-1007

[3] Schwarz DS, Blower MD. The endoplasmic reticulum: Structure, function and response to cellular signaling. Cellular and Molecular Life Sciences. 2016;**73**:79-94

[4] Oslowski CM, Urano F. Measuring ER stress and the unfolded protein response using mammalian tissue culture system. Methods in Enzymology. 2011;**490**:71-92

[5] Wang S, Kaufman RJ. The impact of the unfolded protein response on human disease. The Journal of Cell Biology. 2012;**197**(7):857-867

[6] Hotamisligil GS. Endoplasmic reticulum stress and the inflammatory basis of metabolic disease. Cell. 2010;**140**(6):900-917

[7] Chen Y, Brandizzi F. IRE1: ER stress sensor and cell fate executor. Trends in Cell Biology. 2013;**23**(11):547-555

[8] Casas-Tinto S, Zhang Y, Sanchez-Garcia J, Gomez-Velazquez M, Rincon-Limas DE, Fernandez-Funez P. The ER stress factor XBP1s prevents amyloid-beta neurotoxicity. Human Molecular Genetics. 2011;**20**(11):2144-2160

[9] Kumar V, Maity S. ER stress-sensor proteins and ER-mitochondrial crosstalk-signaling beyond (ER) stress response. Biomolecules. 2021;**11**(2):173

[10] Hetz C. The unfolded protein response: Controlling cell fate decisions under ER stress and beyond. Nature Reviews. Molecular Cell Biology. 2012;**13**(2):89-102

[11] Shen J, Kelleher RJ. The presenilin hypothesis of Alzheimer's disease: Evidence for a loss-of-function pathogenic mechanism. Proceedings of the National Academy of Sciences of the United States of America. 2007;**104**(2):403-409

[12] Dover M, Kishek M, Eddins M, Desar N, Fiala M. The Role of M1- and M2-Type Macrophages in Neurological and Infectious Diseases. London, UK: IntechOpen; 2022. Available from: https://www.intechopen.com/online-first/80219

[13] Schröder M, Kaufman RJ. ER stress and the unfolded protein response. Mutation Research. 2005;**569**(1-2):29-63

[14] Ow SY, Dunstan DE. A brief overview of amyloids and Alzheimer's disease. Protein Science a Publication of Protein Society. 2014;**23**(10):1315-1331

[15] Olivera-Perez HM, Lam L, Dang J, Jiang W, Rodriguez F, Rigali E, et al. Omega-3 fatty acids increase the unfolded protein response and improve amyloid-β phagocytosis by macrophages of patients with mild cognitive impairment. FASEB. 2017;**31**(10):4359-4369

[16] Manoharan S, Guillemin GJ, Abiramasundari RS, Essa MM, Akbar M, Akbar MD. The role of reactive oxygen species in the pathogenesis of Alzheimer's disease, Parkinson's disease, and Huntington's disease: A mini review. Oxidative Medicine and Cellular Longevity. 2016;**2016**:8590578

[17] Llanos-González E, Henares-Chavarino ÁA, Pedrero-Prieto CM, García-Carpintero S, Frontiñán-Rubio J, Sancho-Bielsa FJ, et al. Interplay between mitochondrial oxidative disorders and Proteostasis in Alzheimer's disease. Frontiers in Neuroscience. 2020;**13**:1444

[18] Rottkamp CA, Raina AK, Zhu X, Gaier E, Bush AI, Atwood CS, et al. Redox-active iron mediates amyloid-beta toxicity. Free Radical Biology & Medicine. 2001;**30**(4):447-450

[19] Atlante A, Calissano P, Bobba A, Giannattasio S, Marra E, Passarella S. Glutamate neurotoxicity, oxidative stress and mitochondria. FEBS Letters. 2001;**497**(1):1-5

[20] Jakaria M, Park SY, Haque ME, Karthivashan G, Kim IS, Ganesan P, et al. Neurotoxic agent-induced injury in neurodegenerative disease model: Focus on involvement of glutamate receptors. Frontiers in Molecular Neuroscience. 2018;**11**:307

[21] Lindholm D, Korhonen L, Eriksson O, Kõks S. Recent insights into the role of unfolded protein response in ER stress in health and disease. Frontiers in Cell and Development Biology. 2017;**5**:48

[22] Hutton M, Hardy J. The presenilins and Alzheimer's disease. Human Molecular Genetics. 1997;**6**(10):1639-1646

[23] Katayama T, Imaizumi K, Manabe T, Hitomi J, Kudo T, Tohyama M. Induction of neuronal death by ER stress in Alzheimer's disease. Journal of Chemical Neuroanatomy. 2004;**28**(1-2):67-78

[24] Doyle KM, Kennedy D, Gorman AM, Gupta S, Healy SJM, Samali A. Unfolded proteins and endoplasmic reticulum stress in neurodegenerative disorders. Journal of Cellular and Molecular Medicine. 2011;**15**(10):2025-2039

[25] Fu J, Tao T, Li Z, Chen Y, Li J, Peng L. The roles of ER stress in epilepsy: Molecular mechanisms and therapeutic implications. Biomedicine & Pharmacotherapy. 2020;**1**(131):110658

[26] Falco-Walter JJ, Scheffer IE, Fisher RS. The new definition and classification of seizures and epilepsy. Epilepsy Research. 2018;**139**:73-79

[27] Yamamoto A, Schindler CK, Murphy BM, Bellver-Estelles C, So NK, Taki W, et al. Evidence of tumor necrosis factor receptor 1 signaling in human temporal lobe epilepsy. Experimental Neurology. 2006;**202**(2):410-420

[28] Rao YL, Ganaraja B, Joy T, Pai MM, Ullal SD, Murlimanju BV. Neuroprotective effects of resveratrol in Alzheimer's disease. Frontiers in Bioscience Elite. 2020;**12**(1):139-149

[29] Bando Y, Katayama T, Kasai K, Taniguchi M, Tamatani M, Tohyama M. GRP94 (94 kDa glucose-regulated protein) suppresses ischemic neuronal cell death against ischemia/reperfusion injury. The European Journal of Neuroscience. 2003;**18**(4):829-840

[30] Chen J, Zheng G, Guo H, Shi Z nan, Jiang J, Wang X yu, et al. The effect of metformin treatment on endoplasmic reticulum (ER) stress induced by status epilepticus (SE) via the PERK-eIF2α-CHOP pathway. Bosnian Journal of Basic Medical Sciences 2018;**18**(1):49-54

[31] Engel T, Sanz-Rodgriguez A, Jimenez-Mateos EM, Concannon CG, Jimenez-Pacheco A, Moran C, et al. CHOP regulates the p53-MDM2 axis and is required for neuronal survival after seizures. Brain: A Journal of Neurology. 2013;**136**(Pt 2):577-592

[32] Graves MC, Fiala M, Dinglasan LAV, Liu NQ, Sayre J, Chiappelli F, et al. Inflammation in amyotrophic lateral sclerosis spinal cord and brain is mediated by activated macrophages, mast cells and T cells. Amyotrophic Lateral Sclerosis and Other Motor Neuron Disorder. 2004;**5**(4):213-219

[33] Fiala M, Chattopadhay M, La Cava A, Tse E, Liu G, Lourenco E, et al. IL-17A is increased in the serum and in spinal cord CD8 and mast cells of ALS patients. Journal of Neuroinflammation. 2010;**7**(1):76

[34] Prell T, Stubendorff B, Le TT, Gaur N, Tadić V, Rödiger A, et al. Reaction to endoplasmic reticulum stress via ATF6 in amyotrophic lateral sclerosis deteriorates with aging. Frontiers in Aging Neuroscience. 2019;**11**:5

[35] Bruijn LI, Miller TM, Cleveland DW. Unraveling the mechanisms involved in motor neuron degeneration in ALS. Annual Review of Neuroscience. 2004;**27**:723-749

[36] Rosen DR, Siddique T, Patterson D, Figlewicz DA, Sapp P, Hentati A, et al. Mutations in Cu/Zn superoxide dismutase gene are associated with familial amyotrophic lateral sclerosis. Nature. 1993;**362**(6415):59-62

[37] Youle RJ, Strasser A. The BCL-2 protein family: Opposing activities that mediate cell death. Nature Reviews. Molecular Cell Biology. 2008;**9**(1):47-59

[38] Wootz H, Hansson I, Korhonen L, Näpänkangas U, Lindholm D. Caspase-12 cleavage and increased oxidative stress during motoneuron degeneration in transgenic mouse model of ALS. Biochemical and Biophysical Research Communications. 2004;**322**(1):281-286

[39] Nakagawa T, Zhu H, Morishima N, Li E, Xu J, Yankner BA, et al. Caspase-12 mediates endoplasmic-reticulum-specific apoptosis and cytotoxicity by amyloid-beta. Nature. 2000;**403**(6765):98-103

[40] Fiala M, Popik W, Qiao JH, Lossinsky AS, Alce T, Tran K, et al. HIV-1 induces cardiomyopathyby cardiomyocyte invasion and gp120, tat, and cytokine apoptotic signaling. Cardiovascular Toxicology. 2004;**4**(2):97-107

[41] Twu C, Liu NQ, Popik W, Bukrinsky M, Sayre J, Roberts J, et al. Cardiomyocytes undergo apoptosis in human immunodeficiency virus cardiomyopathy through mitochondrion- and death receptor-controlled pathways. Proceedings of the National Academy of Sciences of the United States of America. 2002;**99**(22):14386-14391

[42] Andhavarapu S, Katuri A, Bryant J, Patel V, Gupta U, Asemu G, et al. Intersecting roles of ER stress, mitochondrial dysfunction, autophagy, and calcium homeostasis in HIV-associated neurocognitive disorder. Journal of Neurovirology. 2020;**26**(5):664-675

[43] Nooka S, Ghorpade A. HIV-1-associated inflammation and antiretroviral therapy regulate astrocyte endoplasmic reticulum stress responses. Cell Death Discovery. 2017;**3**:17061

[44] Fan Y, Gao X, Chen J, Liu Y, He JJ. HIV tat impairs neurogenesis through functioning As a notch ligand and activation of notch signaling pathway. The Journal of Neuroscience. 2016;**36**(44):11362-11373

[45] Begum G, Harvey L, Dixon CE, Sun D. ER stress and effects of DHA as an ER stress inhibitor. Translational Stroke Research. 2013;**4**(6):635-642

[46] Harvey LD, Yin Y, Attarwala IY, Begum G, Deng J, Yan HQ, et al. Administration of DHA reduces endoplasmic reticulum stress-associated inflammation and alters microglial or macrophage activation in traumatic brain injury. ASN Neuro. 2015;7(6):1759091415618969

[47] Hong Y, Wang X, Sun S, Xue G, Li J, Hou Y. Progesterone exerts neuroprotective effects against Aβ-induced neuroinflammation by attenuating ER stress in astrocytes. International Immunopharmacology. 2016;**33**:83-89

Chapter 6

# Advanced Glycation End Product Induced Endothelial Dysfunction through ER Stress: Unravelling the Role of Paraoxonase 2

*Ramya Ravi and Bharathidevi Subramaniam Rajesh*


## Abstract

Hyperglycemia accelerates the formation of advanced glycation end products (AGEs). AGEs are a heterogeneous group of compounds generated by non-enzymatic glycation of proteins or lipids with glucose through Amadori rearrangement and its accumulation increases with aging in diabetes. AGEs augments ROS generation, diminishes the antioxidant defense of the cells, decreases mitochondrial membrane potential, ATP production, and elevates the levels of mitochondrial fission protein (Drp1) and mitophagic proteins (Parkin and PTEN) leading to dysfunction of mitochondria. In this chapter, we have discussed how AGEs trigger the endoplasmic reticulum stress and inflammation and mediate endothelial dysfunction in diabetes and also have discussed the role played by endogenous Paraoxonase 2 (PON2) in mitigating endothelial dysfunction by inhibiting the adverse effects of AGE.


**Keywords:** advanced glycation end product, ER stress, mitochondrial fission, paraoxonase 2

## 1. Introduction

Diabetes mellitus (DM) is a growing metabolic health problem, and around 346 million people are currently affected by diabetes worldwide, and it is anticipated to double by 2030 [http://www.who.int/mediacentre/factsheets/fs312/en/index.html]. Macro and microvascular complications are the major cause of morbidity and mortality in patients with type 2 diabetes [1]. The important pathology associated with vascular diseases is the formation and accumulation of atherosclerotic plaques leading to the process of atherosclerosis, which ultimately results in the narrowing of the blood vessels [2].

Vascular endothelial cells (ECs) are monolayers of cells that line the inner surface of the luminal vessel wall. It acts as a physical barrier between the bloodstream and the luminal wall [3, 4]. The term endothelial cell dysfunction (ECD) is referred to as loss/dysregulation of EC cells function: such as impairment of the barrier function, reduction of the anti-coagulants, disturbance in the balance between the vasodilation and vasoconstrictions, increased pro-inflammatory response, attachment, and adhesion of leukocytes to the endothelial cell surface, increase the production of pro-oxidant molecules [5, 6]. These points highlight the importance of ECs in maintaining normal vascular homeostasis.

During endothelial injury, oxidized lipids and protein adducts (advanced glycation end products) accumulate in the arterial walls. Then the circulating monocytes adhere to the endothelial cells that express adhesion molecules, such as vascular adhesion molecule-1 (VCAM-1) and selectins, and then migrate into the sub-endothelial space. These monocytes that infiltrate the arterial wall get differentiated into macrophages, and this macrophage also accumulates oxidized lipids to form foam cells. Then the foam cells attract the T-lymphocytes, which in turn induce the proliferation of smooth muscle cells in arterial walls. The entire process leads to the formation of a lipid-rich atherosclerotic lesion and rupture of this lesion leads to vascular infarction. Additionally, increased platelet aggregation and coagulation are also observed due to impaired nitric oxide generation, free-radical generation from the platelet, and elevated levels of plasminogen activator inhibitor. All these factors contribute to the vascular wall occlusion and further increase the risk of cardiovascular events as shown in **Figure 1** [2, 7].

Numerous studies have demonstrated the existence of an association between elevated AGE levels and cardiovascular disease in DM patients. In line with this, clinical studies reported that CML and pentosidine levels are increased in the progression of the disease and act as a predictor of cardiovascular events [8–14]. A fourfold increase in the incidence of coronary artery disease, a 10-fold increase in peripheral vascular disease and a 3–4-fold higher mortality rate with as much as 75% of diabetics ultimately dying from vascular disease have been reported [15].

Vascular complications are classified into two types, namely macrovascular and microvascular. Wherein the large vessels such as arteries and veins get affected in macrovascular disease and microvascular involves small vessels such as capillaries. Chronic hyperglycemia initiates the production of AGE and turns on diabetic vascular complications through elevated production of reactive oxygen species, which

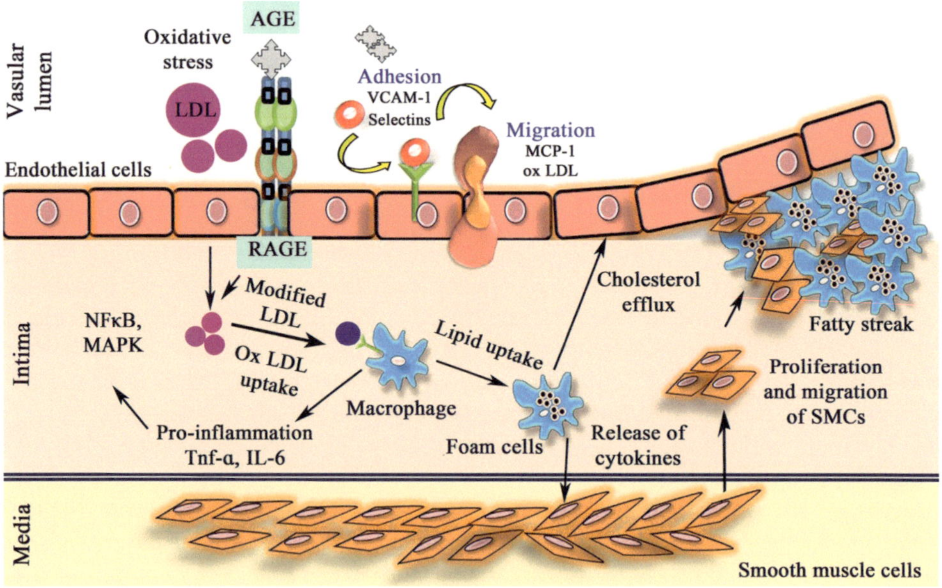


**Figure 1.**
*Mechanism of AGEs-induced oxidative stress in the formation of atherosclerotic plaque.*

instigates various signaling cascades such as Receptor for advanced glycation end product receptor (RAGE) activation, protein kinase C (PKC) and Mitogen activated protein kinases (MAPK) pathways. Macrovascular complications associated with diabetes are arteriosclerotic cardiovascular diseases (ASCVDs) such as coronary heart disease (CHD), peripheral artery disease (PAD), and stroke [2]. Microvascular complications associated with diabetes are diabetic retinopathy (DR), neuropathy, and nephropathy [16, 17] as shown in **Figure 2**.

The pathogenic mechanisms underlying diabetic nephropathy involve the generation of reactive oxygen species (ROS), accumulation of AGEs, and activation of intracellular signaling molecules such as PKC [18, 19]. Studies suggest that increased formation of AGEs in the vitreous may be involved in the development of diabetic retinopathy by inducing the production of interleukins –6 (IL-6) from retinal Müller cells [20]. AGEs interaction RAGE plays an important role in the pathogenesis of DR [21]. In DR, AGEs (both early and late Amadori products) have been localized to vascular cells (endothelial and pericytes), neurons, glia, and also in the vitreous [22, 23]. Pentosidine and Nε-(carboxymethyl)lysine (CML) levels are increased in aqueous, vitreous, and serum of DR patients when compared with non-diabetic controls especially increased in proliferative DR (PDR) patients than non-proliferative DR (NPDR) and referred to as biomarker for microvascular complications [24, 25] with the progression of DR with decreased visual acuity, emphasizing that AGEs are novel biomarkers/risk markers for type 2 DR [26, 27]. In addition to the above AGEs, methylglyoxal derivative hydroimidazolone was also increased in DR patients [28, 29].

AGEs are a heterogeneous group of compounds generated by non-enzymatic glycation of proteins or lipids with glucose through Amadori rearrangement and

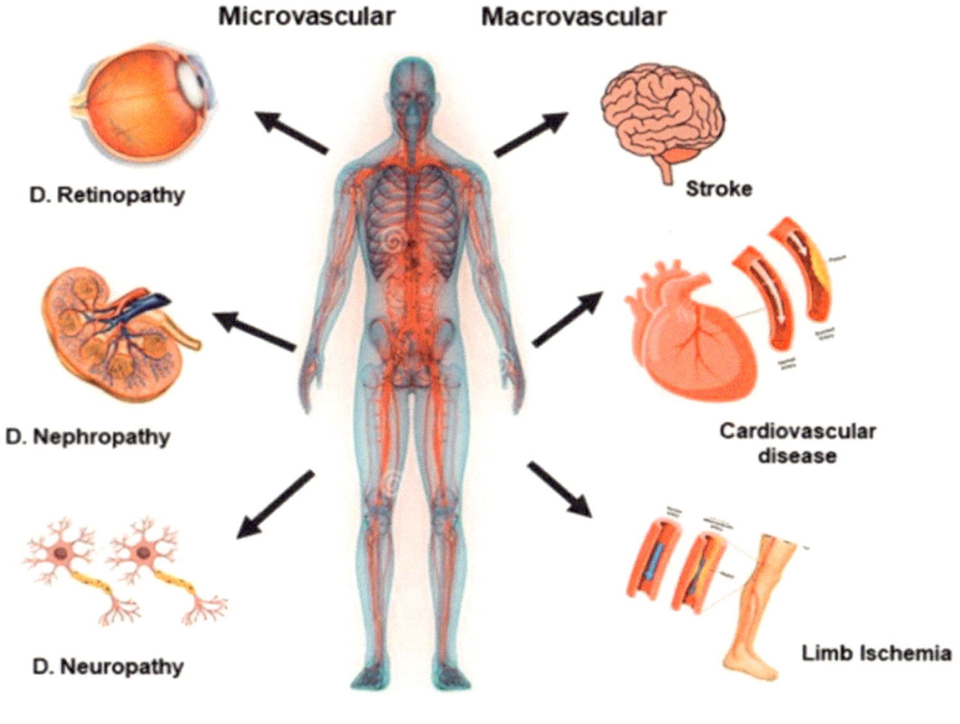


**Figure 2.**
*AGEs-induced macrovascular and microvascular complications.*

its accumulation increases with aging and in diabetes [30]. The term AGEs has been applied to a broad range of advanced glycation end products such as CML, $N^{\varepsilon}$-(carboxymethyl)hydroxylysine, pyrraline, and pentosidine [31]. Among these, CML is the predominant epitope in the AGE adducts detected in tissue proteins of diabetic patients [32–34]. Its level is found to be increased in serum and aqueous humor of type 2 diabetic patients with retinopathy (DR) [2]. It is also used as a biomarker to predict the progression of different stages of DR. AGEs-stimulated cell response is initiated by its engagement with the receptors present on the cell surface [35]. The most studied AGE receptor is RAGE. Several other AGE receptors identified so far consist of AGE-receptor complex (AGER1/OST-48, AGER2/80 K-H, AGER3/Gal-3), some members of the Toll-like Receptor (TLRs) family (TLR4, TLR2), and scavenging receptors (SRs) family (SR-AI, SR-AII, CD36, LOX-1, FEEL-1, and FEEL-2). The expression of these AGE-activated receptors depends on the cell/tissue type [36].

RAGE and TLRs are well-known Pattern Recognition Receptors (PRRs), which recognize molecules found in pathogens (Pathogen-Associated Molecular Patterns–PAMPs, ex-LPS) or molecules released from damaged or stressed cells such as high mobility group B (HMGB1) and serum amyloid proteins referred as Damage-Associated Molecular patterns (DAMPs) [37]. It is expressed on the surface of various cells, including endothelial, epithelial, and fibroblast mediates multiple signaling pathways such as MAPK kinase and Nuclear factor kappa B (NFκB), which activate a pro-inflammatory response [38, 39]. Among various TLRs, TLR-2 and 4 are reported to be increased in monocytes of type 1 and type 2 diabetic patients [40, 41] associated with microvascular complications such as retinopathy, nephropathy, and neuropathy [42].

Although various pathways are involved in the pathogenesis of endothelial dysfunction such as activation of the polyol pathway, auto-oxidation of glucose, PKC pathway activation, and formation of AGEs, all these pathways may intersect at several points to increase the complexity of the disease [43]. Among these, increased formations of AGEs (advanced glycation end product) is one of the causes of ECs dysfunction, which has been implicated in the pathogenesis of diabetes-induced vascular complications, as evident by their *in vivo* accumulation as reported in numerous diseases [44]. In this chapter, we would focus on how AGEs induce an inflammatory response, mitochondrial dysfunction, and endothelial dysfunction through ER stress.

## 2. AGE-induced effect in endothelial cells

The high glucose-induced “oxidative stress” and “endoplasmic reticulum (ER) stress” of the endothelium may play major roles in the initiation and progression of cardiovascular clinical manifestations in diabetes [45]. While diabetes management has largely focused on the control of hyperglycemia, the rising burden of this disease is mainly correlated to its vascular complications [46]. This is reflected by a type II diabetes differs principally from type I diabetes in that it is accompanied by a period of hyperinsulinemia and is characterized by late as opposed to early onset of hyperglycemia. In type I DM, vascular involvement (through endothelial dysfunction) occurs as a result of metabolic insult/hyperglycemia, while in type II DM, endothelial dysfunction plays a more direct role and is aggravated by, rather than caused by, hyperglycemia [6].

### 2.1 AGE and oxidative stress

One of the best-characterized actions of AGEs on ECs is the induction of ROS [47, 48]. There is considerable evidence to show that AGE induces ROS generation and diminishes the antioxidant defense of the cells [49–51].

Oxidative stress is induced either by the abundant production of ROS or the failure of the antioxidative machinery mechanism. The main source of ROS is the electrons present in the mitochondrial respiratory chain, which results in the formation of superoxide anion (O2-), hydroxyl radical (OH.) hydrogen peroxide ($H_2O_2$). AGEs induce mitochondrial dysfunction through the generation of ROS production.

Mitochondria are complex organelle that undergoes complete fusion and division under physiological or pathological conditions. Mitochondrial fission is defined as the division of a mitochondrion from two or more separate mitochondrial compartments, and this process is essential in the distribution of mtDNA during cell division and helps in the removal of damaged mitochondria through mitophagy. Mitochondrial fusion is merging two or more mitochondria. It is regulated by at least three proteins: optic atrophy 1 (OPA1), mitofusin 1 (MFN1), and mitofusin 2 (MFN2), and mitochondrial fission are controlled by dynamin-related protein 1 (DRP1/DLP1/DNM1), fission 1 (FIS1), and mitochondrial fission factor (MFF) [52, 53]. Studies show that in diabetes, mitochondrial fission is increased [54–56]. Diabetic animal model (diabetic-STZ model) study demonstrated that HG alters the mitochondrial respiration and alters glomerular bioenergetics, whereas podocytes isolated from those mice showed fragmented mitochondria [57]. Another study in Sprague-Dawley rats (STZ induced) reported an increase in neuronal pyknosis with increased DRP1 expression in neurons when compared with Normal control (NC) group associating that High glucose (HG) aggravated ischemic brain damage and alteration in mitochondrial dynamics [58]. Very importantly, AGEs such as AGE-BSA promoted mitochondrial fission, loss of membrane potential, and apoptosis through the RAGE pathway, whereas blocking the RAGE signaling reduces the events of mitochondrial abnormalities in endothelial and osteoblastic cells [59, 60]. In a rodent model, AGE-BSA disturbs the mitochondrial respiratory chain and induces mitochondrial pore formation with an increase in RAGE expression suggesting the fact that AGE-induced mitochondrial dysfunction plays a major role in diabetic neuropathy [61]. Proteomics study by tandem mass spectrometry (nanoLC-ESI-ETD MS/MS) revealed that renal tubular cells exposed to HG increased phosphorylation and oxidation of mitochondrial proteins when compared with the NG [62]. CML-BSA, another AGE, produces detrimental effects in diabetic db/db mice inducing mitophagy. In this study, CML-BSA treatment in pancreatic β-cells increased RAGE and ROS with a decrease in membrane potential and ATP production with elevated levels of mitochondrial fission (Drp1) and mitophagic proteins (Parkin and PTEN) supporting the factor that increases the concentration of AGEs damage the β-cells and reduces the insulin cells function by making vulnerable to AGEs-induced damages [63].

Under physiological conditions, low concentrations of ROS are needed to maintain cellular proliferation, migration, and survival [64]. However, during the pathological condition, overproduction of ROS induces deleterious effects on the cells and tissues, and to overcome this, the cellular environment has an antioxidative defense system that is capable of scavenging the ROS. The antioxidant defense system includes low-molecular ROS scavengers, antioxidative enzymes, and degrading or repairing proteins [50]. The low-molecular ROS scavengers include glutathione (GSH), vitamin C, and D. Secondly, the antioxidative enzymes superoxide dismutases (SOD) and catalase, which can convert ROS into less reactive ions. The last part of the defense system includes the proteasome and protease system, which will degrade the damaged proteins. The impairments in the antioxidant defense system lead to oxidative stress, which eventually activates various cellular signaling pathways.

### 2.2 Pro-inflammatory response by AGE

Further, AGEs also increase pro-inflammatory response, by augmenting endoplasmic reticulum (ER) stress and amplifying the angiogenic potential of ECs, which has been observed in a myriad of human diseases such as atherosclerosis, acute/chronic inflammatory diseases, vascular complication, and aging [65–68].

Numerous AGEs-induced vascular diseases display elevated levels of pro-inflammatory cytokine either by directly activating ROS or through engaging with RAGE receptor. AGE-RAGE is a prominent axis that facilitates the activation of NFκB and MAPK signaling pathways, which subsequently induces the expression of cytokines, chemokines, and adhesion molecules [38, 69]. Elevated levels of AGEs enhanced the release of pro-inflammatory cytokines such as interleukin-6 (IL-6), interleukin-8 (IL-8), tumor necrosis factor (TNF-α), and interferon-beta (IL-1β). RAGE also binds HMGB1, commonly referred to as damage-associated molecules (DAMPs), and acts as an inflammatory mediator and is released into circulation when there is an injury [70]. Recognition of HMGB1 by RAGE induces the NFκB activation and its downstream cytokines IL-6, IL-8, and TNF-α. [70, 71] Accumulating evidence shows that DAMPs are reported to mediate the pathogenesis of atherosclerosis and diabetic vascular complications.

Another important feature of inflammatory response is the recruitment and attachment of circulating leukocytes to the endothelium, which is the initial stage in the development of atherosclerosis. Increased adhesion of leukocytes to ECs is facilitated by enhanced expression of chemokines such as monocyte chemoattractant protein (MCP-1) and adhesion molecules such as intercellular adhesion molecules (ICAM), vascular cell adhesion molecules (VCAM), and E-selectins. Studies also demonstrate that direct interaction exists between the ECs adhesion molecules and lymphocyte-function-associated antigen 1 (LFA-1), macrophage-1 antigen (Mac-1), and very late antigen-1 (VLA-1), which are the major counter receptors present on the surface of the leukocytes [39, 72]. Abundant studies show that AGEs elevate the expression of adhesion molecules on the surface of the endothelial cells [73, 74].

### 2.3 AGE increases ER stress

Numerous studies provided molecular insights, highlighting the functional link existing between endothelial dysfunction and endoplasmic reticulum (ER) stress [75]. ER is the vital organelle that plays important role in protein folding, lipid biosynthesis, and calcium ($Ca^{2+}$) regulation. When the ER homeostasis is disturbed either due to increased synthesis of protein, accumulation of misfolded proteins, increased oxidative stress, or alteration in calcium load leads to a condition termed as ER stress. It is closely monitored by the evolutionarily conserved quality control system called unfolded protein response (UPR) [75, 76].

The UPR system is activated by three transmembrane proteins: 1) RNA-dependent protein kinase-like ER eukaryotic initiation factor-2α kinase (PERK), 2) inositol-requiring ER-to-nucleus signaling protein 1 (IRE1), and 3) activating transcription factor 6 (ATF6). These three sensors are inactive form when they are bound with glucose-regulated protein kinase-78 (GRP78) and released upon induction of ER stress. These sensors resolve the ER stress by 1) activating the induction of UPR genes, which enhances the protein folding; 2) attenuation of protein translation, therefore, it reduces the workload of ER; and 3) activating ER-associated degradation (ERAD) pathway activation, which eliminates the unfolded protein through proteasome degradation pathway [75, 77].

During ER stress, activation of the PERK pathway leads to phosphorylation of the eukaryotic translation initiation factor 2 alpha (eIF2α), resulting in the attenuation of protein translation and activating the ATF4 (activating transcription factor-4). ATF4 subsequently induces the activation of pro-apoptotic mediator CHOP (C/EBPα-homologous protein, also known as GADD153) and its downstream target gene, DNA-damage-inducible protein-34 (GADD34). Activation of IRE1 promotes splicing of the endoribonuclease unconventional splicing of an mRNA encoding the transcription factor X-box-binding protein 1 (XBP1s). XBP1s activation further induces the activation of UPR genes, which either increases the ER folding capacity or intersects with AFT4-CHOP to activate the apoptosis. Another UPR pathway is the activation of AFT6, which translocates to the Golgi apparatus. It alleviates misfolded proteins through activation of the ERAD pathway mediated by ATF6-XBP1as shown in **Figure 3** [75, 76].

Studies reported that under stress conditions, ER show an unusual morphological pattern called ER whorls, which can be used as a biomarker for ER stress [78]. Further compared with the normal cells, DTT treatment (inducer of ER stress) activated the formation of ER whorls, which are seen beneath the plasma membrane, which is in turn imported into the vacuole and proceeds for autophagy [79]. Earlier a study has shown the presence of whorl-like sER in the Leydig cells of STZ-induced mouse model indicating that diabetes can induce morphological changes in ER. Till now there are no studies that have described the effect of AGE on ER morphology. Whereas there are studies that have highlighted the distorted mitochondrial

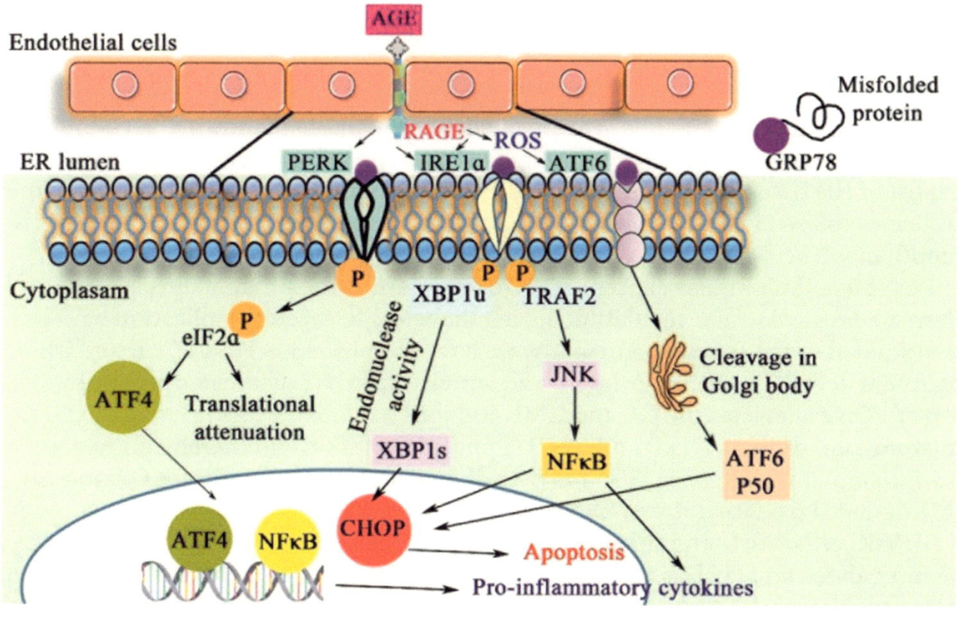


**Figure 3.**
*AGEs-induced ER stress in endothelial cells.*

morphology (mitochondrial fission) seen during mitochondrial fission in diabetic animal models [80, 81].

ER and mitochondria are connected through mitochondrial- associated ER membranes (MAMs), which help in the transfer of $Ca^{2+}$, ATP, and metabolites [82, 83]. As ER stress magnitudes, it increases the releases of $Ca^{2+}$ from ER to mitochondria, which leads to the opening of mitochondrial membrane pore and releases of cytochrome C, which further intensifies the ROS production. Taken together, it creates a vicious cycle of ER stress and mitochondrial dysfunction, which progresses toward apoptotic signaling [84].

A vast number of studies link ER stress with inflammatory and oxidative signaling pathways, which play a putative role in the development and progression of endothelial dysfunction. Several lines of evidence support the fact that ER stress acts as a potent inflammatory activator because each of the UPR arms activates NFκB pathway, which in turn releases an array of inflammatory cytokines IL-6, IL-8, and TNF-α [85, 86]. A lot of *in vivo* and *in vitro* studies also support the notion that AGEs-induced ER stress promotes the activation of ROS, autophagy, pro-inflammatory cytokines response, and apoptosis in most of diabetic vascular complications [85, 87–91].

Although several studies have identified an array of molecular entities and pathways that activate endothelial dysfunction, the cellular processes underlying endothelial dysfunction are majorly oxidative stress and inflammation. Increasing the intake of fruits and vegetables rich in polyphenols and flavonoids (anti-inflammatory and antioxidant properties) protects the endothelium and reduces the risk of cardiovascular complications [92, 93]. Increasing the endogenous antioxidant may pay way to develop a more effective and safer option. One such endogenous molecule with antioxidative, anti-apoptotic, and anti-inflammatory properties is paraoxonase [94]. Decreased serum paraoxonase (PON) activity is seen in both diabetes, and its complications have been reported, which is attributed to its glycation [95, 96]. The human PON enzyme consists of three family members, paraoxonase (PON1), paraoxonase 2 (PON2), and paraoxonase 3 (PON3). These genes are located on the long arm of chromosome no 7 (7q21-22) with nine exons. Based on the structural homology and from the evolutionary point of view, PON2 is reported to be the oldest member of the family followed by PON3 and PON1 [97]. PON1 is HDL associated and secreted in serum, whereas PON2 and PON3 are located intracellularly mainly in the endoplasmic reticulum, mitochondria, and the nuclear membrane and are ubiquitously expressed in most of the tissues including the liver, kidney, intestine, placenta, etc. [98]. Its role has been explored in many cells including epithelial, endothelial, macrophages, and smooth muscle cells.

PON2 is a ubiquitously expressed antioxidant and anti-inflammatory protein, where its expression and regulation during diabetes- induced complication have not been studied so far. In our recent study, we have established in HUVECs that AGE treatment decreases mRNA, protein, and activity of PON2, whereas overexpression of PON2 alleviates the GA and CML-induced oxidative stress, ER stress, and inflammation through NFκB and ERK1/2 phosphorylation and thereby mitigates pro-inflammatory response [99]. Further silencing of PON2 aggravates GA and CML-induced oxidative stress, ER stress, and pro-inflammatory cytokines expression in HUVEC cells. We found that in diabetic retina PON2 expression was significantly downregulated and HRECs treatment with CML increased mitochondrial fission and aggravates the mitochondrial-dependent apoptosis [100]. Conversely, overexpression of PON2 inhibits the JNK1/2-mediated signaling pathway and rescues the cells from mitochondrial fission and apoptosis as shown in **Figure 4** [100].

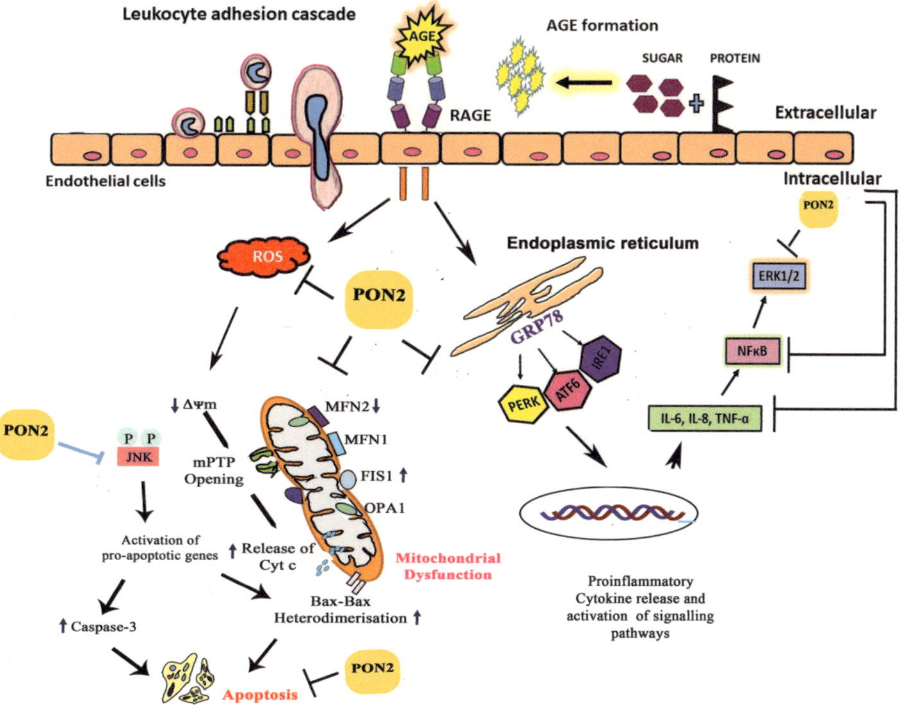


**Figure 4.**
*Possible mechanistic role of PON2 in mitigating AGEs-induced ER stress, pro-inflammation, and mitochondrial dysfunction in endothelial cells.*

## 3. Conclusion

There are a large number of studies that have established the deleterious effects of AGE on various cellular models. In this chapter, we have discussed in detail how AGEs induce mitochondrial dysfunction, inflammation and endothelial dysfunction through augmenting ER stress. We have also highlighted the association between mitochondrial stress and ER stress. The role of antioxidant PON in inhibiting these deleterious effects has also been discussed.

## Acknowledgements

We thank the Council of Scientific and Industrial Research (CSIR-27 (0310)/14-EMR-11) and Indian Council of Medical Research (ICMR-ID no: 2017-0976/CMB-BMS) for providing the funding and the fellowship.

## Conflict of interest

The authors declare no conflict of interest.

## Author details

Ramya Ravi and Bharathidevi Subramaniam Rajesh*
R.S. Mehta Jain Department of Biochemistry and Cell Biology, KBIRVO, Vision Research Foundation, Chennai, India

*Address all correspondence to: drbarathi@snmail.org

## References

[1] Beckman JA, Creager MA. Vascular complications of diabetes. Circulation Research. 2016;**118**:1771-1785

[2] Fowler MJ. Microvascular and macrovascular complications of diabetes. Clinical Diabetes. 2008;**26**:77-82

[3] Sena CM, Pereira AM, Seiça R. Endothelial dysfunction - a major mediator of diabetic vascular disease. Biochimica et Biophysica Acta. 1832;**2013**:2216-2231

[4] Kolluru GK, Bir SC, Kevil CG. Endothelial dysfunction and diabetes. International Journal of Vascular Medicine. 2012;**2012**:918267

[5] Deanfield JE, Halcox JP, Rabelink TJ. Endothelial function and dysfunction. Circulation. 2007;**115**:1285-1295

[6] Basha B, Samuel SM, Triggle CR, Ding H. Endothelial dysfunction in diabetes mellitus. Experimental Diabetes Research. 2012;**2012**:481840

[7] Libby P, Buring JE, Badimon L, Hansson GK, Deanfield J, Bittencourt MS, et al. Atherosclerosis. Nature Reviews. Disease Primers. 2019;**5**:56

[8] Basta G, Berti S, Cocci F, Lazzerini G, Parri S, Papa A, et al. Plasma N-epsilon-(carboxymethyl)lysine levels are associated with the extent of vessel injury after coronary arterial stenting. Coronary Artery Disease. 2008;**19**:299-305

[9] Hanssen NMJ, Engelen L, Ferreira I, Scheijen JLJM, Huijberts MS, van Greevenbroek MMJ, et al. Plasma levels of advanced glycation endproducts Nε-(carboxymethyl)lysine, Nε-(carboxyethyl)lysine, and pentosidine are not independently associated with cardiovascular disease in individuals with or without type 2 diabetes. The Journal of Clinical Endocrinology and Metabolism. 2013;**98**:E1369-E1373

[10] Semba RD, Bandinelli S, Sun K, Guralnik JM, Ferrucci L. Plasma carboxymethyl-lysine, an advanced glycation end product, and all-cause and cardiovascular disease mortality in older community-dwelling adults. Journal of the American Geriatrics Society. 2009;**57**:1874-1880

[11] Busch M, Franke S, Stein G, Wolf G. Ist die Serumkonzentration von Pentosidin ein Prädiktor kardiovaskulärer Ereignisse bei Patienten mit Diabetes mellitus Typ 2 und Nephropathie? Deutsche Medizinische Wochenschrift. 1946;**132**(2007):1810-1814

[12] de Vos LC, Lefrandt JD, Dullaart RPF, Zeebregts CJ, Smit AJ. Advanced glycation end products. Atherosclerosis. 2016;**254**:291-299

[13] Indyk D, Bronowicka-Szydełko A, Gamian A, Kuzan A. Advanced glycation end products and their receptors in serum of patients with type 2 diabetes. Scientific Reports. 2021;**11**:13264

[14] Li J, Shangguan H, Chen X, Ye X, Zhong B, Chen P, et al. Advanced glycation end product levels were correlated with inflammation and carotid atherosclerosis in type 2 diabetes patients. Open Life Sciences. 2020;**15**:364-372

[15] Grundy SM, Howard B, Smith S, Eckel R, Redberg R, Bonow RO.

Prevention conference VI. Circulation. 2002;**105**:2231-2239

[16] Chawla A, Chawla R, Jaggi S. Microvasular and macrovascular complications in diabetes mellitus. Indian Journal of Endocrinology and Metabolism. 2016;**20**:546-551

[17] Paul S, Ali A, Katare R. Molecular complexities underlying the vascular complications of diabetes mellitus - a comprehensive review. Journal of Diabetes and its Complications. 2020;**34**:107613

[18] Trevisan R, Dodesini AR, Lepore G. Lipids and renal disease. Journal of the American Society of Nephrology JASN. 2006;**17**:S145-S147

[19] Cade WT. Diabetes-related microvascular and macrovascular diseases in the physical therapy setting. Physical Therapy. 2008;**88**:1322-1335

[20] Nakamura N, Hasegawa G, Obayashi H, Yamazaki M, Ogata M, Nakano K, et al. Increased concentration of pentosidine, an advanced glycation end product, and interleukin-6 in the vitreous of patients with proliferative diabetic retinopathy. Diabetes Research and Clinical Practice. 2003;**61**:93-101

[21] Al-Mesallamy HO, Hammad LN, El-Mamoun TA, Khalil BM. Role of advanced glycation end product receptors in the pathogenesis of diabetic retinopathy. Journal of Diabetes and its Complications. 2011;**25**:168-174

[22] Stitt AW. AGEs and diabetic retinopathy. Investigative Ophthalmology & Visual Science. 2010;**51**:4867-4874

[23] Takayanagi Y, Yamanaka M, Fujihara J, Matsuoka Y, Gohto Y, Obana A, Tanito M. Evaluation of Relevance between Advanced Glycation End Products and Diabetic Retinopathy Stages Using Skin Autofluorescence. Antioxidants. Basel. 2020 Nov 9;**9**(11):1100

[24] Kerkeni M, Saïdi A, Bouzidi H, Letaief A, Ben Yahia S, Hammami M. Pentosidine as a biomarker for microvascular complications in type 2 diabetic patients. Diabetes & Vascular Disease Research. 2013;**10**:239-245

[25] Kerkeni M, Saïdi A, Bouzidi H, Ben Yahya S, Hammami M. Elevated serum levels of AGEs, sRAGE, and pentosidine in Tunisian patients with severity of diabetic retinopathy. Microvascular Research. 2012;**84**:378-383

[26] Ghanem AA, Elewa A, Arafa LF. Pentosidine and N-carboxymethyl-lysine. European Journal of Ophthalmology. 2011;**21**:48-54

[27] Choudhuri S, Dutta D, Sen A, Chowdhury IH, Mitra B, Mondal LK, et al. Role of N-ε- carboxy methyl lysine, advanced glycation end products and reactive oxygen species for the development of nonproliferative and proliferative retinopathy in type 2 diabetes mellitus. Molecular Vision. 2013;**19**:100-113

[28] Lapolla A, Flamini R, Dalla Vedova A, Senesi A, Reitano R, Fedele D, et al. Glyoxal and methylglyoxal levels in diabetic patients. Clinical Chemistry and Laboratory Medicine. 2003;**41**: 1166-1173

[29] Fosmark DS, Torjesen PA, Kilhovd BK, Berg TJ, Sandvik L, Hanssen KF, et al. Increased serum levels of the specific advanced glycation end product methylglyoxal-derived hydroimidazolone are associated with retinopathy in patients with type 2

diabetes mellitus. Metabolism: Clinical and Experimental. 2006;**55**:232-236

[30] Chaudhuri J, Bains Y, Guha S, Kahn A, Hall D, Bose N, et al. The role of advanced glycation end products in aging and metabolic diseases. Cell Metabolism. 2018;**28**:337-352

[31] Gkogkolou P, Böhm M. Advanced glycation end products. Dermato-Endocrinology. 2012;**4**:259-270

[32] Ikeda K, Higashi T, Sano H, Jinnouchi Y, Yoshida M, Araki T, et al. N (epsilon)-(carboxymethyl)lysine protein adduct is a major immunological epitope in proteins modified with advanced glycation end products of the Maillard reaction. Biochemistry. 1996;**35**:8075-8083

[33] Murata T, Nagai R, Ishibashi T, Inomuta H, Ikeda K, Horiuchi S. The relationship between accumulation of advanced glycation end products and expression of vascular endothelial growth factor in human diabetic retinas. Diabetologia. 1997;**40**:764-769

[34] Reddy S, Bichler J, Wells-Knecht KJ, Thorpe SR, Baynes JW. N epsilon-(carboxymethyl)lysine is a dominant advanced glycation end product (AGE) antigen in tissue proteins. Biochemistry. 1995;**34**:10872-10878

[35] Shen CY, Lu CH, Wu CH, Li KJ, Kuo YM, Hsieh SC, Yu CL. The Development of Maillard Reaction, and Advanced Glycation End Product (AGE)-Receptor for AGE (RAGE) Signaling Inhibitors as Novel Therapeutic Strategies for Patients with AGE-Related Diseases. Molecules. 2020 Nov 27;**25**(23):5591

[36] Ott C, Jacobs K, Haucke E, Navarrete Santos A, Grune T, Simm A. Role of advanced glycation end products in cellular signaling. Redox Biology. 2014;**2**:411-429

[37] Schmidt AM, Yan SD, Yan SF, Stern DM. The multiligand receptor RAGE as a progression factor amplifying immune and inflammatory responses. The Journal of Clinical Investigation. 2001;**108**:949-955

[38] Bongarzone S, Savickas V, Luzi F, Gee AD. Targeting the receptor for advanced glycation Endproducts (RAGE). Journal of Medicinal Chemistry. 2017;**60**:7213-7232

[39] Kierdorf K, Fritz G. RAGE regulation and signaling in inflammation and beyond. Journal of Leukocyte Biology. 2013;**94**:55-68

[40] Dasu MR, Devaraj S, Park S, Jialal I. Increased toll-like receptor (TLR) activation and TLR ligands in recently diagnosed type 2 diabetic subjects. Diabetes Care. 2010;**33**:861-868

[41] Devaraj S, Dasu MR, Rockwood J, Winter W, Griffen SC, Jialal I. Increased toll-like receptor (TLR) 2 and TLR4 expression in monocytes from patients with type 1 diabetes. The Journal of Clinical Endocrinology and Metabolism. 2008;**93**:578-583

[42] Devaraj S, Jialal I, Yun J-M, Bremer A. Demonstration of increased toll-like receptor 2 and toll-like receptor 4 expression in monocytes of type 1 diabetes mellitus patients with microvascular complications. Metabolism: Clinical and Experimental. 2011;**60**:256-259

[43] de Vriese AS, Verbeuren TJ, van de Voorde J, Lameire NH, Vanhoutte PM. Endothelial dysfunction in diabetes. British Journal of Pharmacology. 2000;**130**:963-974

[44] Tan KCB, Chow W-S, Ai VHG, Metz C, Bucala R, Lam KSL. Advanced glycation end products and endothelial dysfunction in type 2 diabetes. Diabetes Care. 2002;**25**:1055-1059

[45] Zhou Y, Murugan DD, Khan H, Huang Y, Cheang WS. Roles and Therapeutic Implications of Endoplasmic Reticulum Stress and Oxidative Stress in Cardiovascular Diseases. Antioxidants. Basel. 2021 Jul 22;**10**(8):1167

[46] Rodríguez-Gutiérrez R, Montori VM. Glycemic control for patients with type 2 diabetes mellitus, circulation. Cardiovascular Quality and Outcomes. 2016;**9**:504-512

[47] Ramya R, Coral K, Bharathidevi SR. RAGE silencing deters CML-AGE induced inflammation and TLR4 expression in endothelial cells. Experimental Eye Research. 2021;**206**:108519

[48] Ishibashi Y, Matsui T, Isami F, Abe Y, Sakaguchi T, Higashimoto Y, et al. N-butanol extracts of Morinda citrifolia suppress advanced glycation end products (AGE)-induced inflammatory reactions in endothelial cells through its anti-oxidative properties. BMC Complementary and Alternative Medicine. 2017;**17**:137

[49] Poljsak B, Šuput D, Milisav I. Achieving the balance between ROS and antioxidants. Oxidative Medicine and Cellular Longevity. 2013;**2013**:956792

[50] Nowotny K, Jung T, Höhn A, Weber D, Grune T. Advanced glycation end products and oxidative stress in type 2 diabetes mellitus. Biomolecules. 2015;**5**:194-222

[51] Ottum MS, Mistry AM. Advanced glycation end-products. Journal of Clinical Biochemistry and Nutrition. 2015;**57**:1-12

[52] Kluge MA, Fetterman JL, Vita JA. Mitochondria and endothelial function. Circulation Research. 2013;**112**:1171-1188

[53] Davidson SM, Duchen MR. Endothelial mitochondria. Circulation Research. 2007;**100**:1128-1141

[54] Hu L, Ding M, Tang D, Gao E, Li C, Wang K, et al. Targeting mitochondrial dynamics by regulating Mfn2 for therapeutic intervention in diabetic cardiomyopathy. Theranostics. 2019;**9**:3687-3706

[55] Wu Q-R, Zheng D-L, Liu P-M, Yang H, Li L-A, Kuang S-J, et al. High glucose induces Drp1-mediated mitochondrial fission via the Orai1 calcium channel to participate in diabetic cardiomyocyte hypertrophy. Cell Death & Disease. 2021;**12**:216

[56] Kowluru RA, Mohammad G. Epigenetics and mitochondrial stability in the metabolic memory phenomenon associated with continued progression of diabetic retinopathy. Scientific Reports. 2020;**10**:6655

[57] Audzeyenka I, Rachubik P, Typiak M, Kulesza T, Topolewska A, Rogacka D, et al. Hyperglycemia alters mitochondrial respiration efficiency and mitophagy in human podocytes. Experimental Cell Research. 2021;**407**:112758

[58] Liu W-J, Jiang H-F, Rehman FU, Zhang J-W, Chang Y, Jing L, et al. Lycium Barbarum polysaccharides decrease hyperglycemia-aggravated ischemic brain injury through maintaining mitochondrial fission and fusion balance. International Journal of Biological Sciences. 2017;**13**:901-910

[59] Zhang S, Gao Y, Wang JA. Advanced glycation end products influence

mitochondrial fusion-fission dynamics through RAGE in human aortic endothelial cells. International Journal of Clinical and Experimental Pathology. 2017;**10**:8010-8022

[60] Mao YX, Cai WJ, Sun XY, Dai PP, Li XM, Wang Q, et al. RAGE-dependent mitochondria pathway. Cell Death & Disease. 2018;**9**:674

[61] Coughlan MT, Thorburn DR, Penfold SA, Laskowski A, Harcourt BE, Sourris KC, et al. RAGE-induced cytosolic ROS promote mitochondrial superoxide generation in diabetes. Journal of the American Society of Nephrology JASN. 2009;**20**:742-752

[62] Aluksanasuwan S, Plumworasawat S, Malaitad T, Chaiyarit S, Thongboonkerd V. High glucose induces phosphorylation and oxidation of mitochondrial proteins in renal tubular cells. Scientific Reports. 2020;**10**:5843

[63] Lo M-C, Chen M-H, Lee W-S, Lu C-I, Chang C-R, Kao S-H, et al. Nε-(carboxymethyl) lysine-induced mitochondrial fission and mitophagy cause decreased insulin secretion from β-cells, American journal of physiology. Endocrinology and Metabolism. 2015;**309**:E829-E839

[64] Dröge W. Free radicals in the physiological control of cell function. Physiological Reviews. 2002;**82**:47-95

[65] Moldogazieva NT, Mokhosoev IM, Mel'nikova TI, Porozov YB, Terentiev AA. Oxidative stress and advanced Lipoxidation and glycation end products (ALEs and AGEs) in aging and age-related diseases. Oxidative Medicine and Cellular Longevity. 2019;**2019**:3085756

[66] Perillo B, Di Donato M, Pezone A, Di Zazzo E, Giovannelli P, Galasso G, et al. ROS in cancer therapy. Experimental & Molecular Medicine. 2020;**52**:192-203

[67] Snezhkina AV, Kudryavtseva AV, Kardymon OL, Savvateeva MV, Melnikova NV, Krasnov GS, et al. ROS generation and antioxidant defense Systems in Normal and Malignant Cells. Oxidative Medicine and Cellular Longevity. 2019;**2019**:6175804

[68] Nita M, Grzybowski A. The role of the reactive oxygen species and oxidative stress in the Pathomechanism of the age-related ocular diseases and other pathologies of the anterior and posterior eye segments in adults. Oxidative Medicine and Cellular Longevity. 2016;**2016**:3164734

[69] Teissier T, Boulanger É. The receptor for advanced glycation end-products (RAGE) is an important pattern recognition receptor (PRR) for inflammaging. Biogerontology. 2019;**20**:279-301

[70] Steinle JJ. Role of HMGB1 signaling in the inflammatory process in diabetic retinopathy. Cellular Signalling. 2020;**73**:109687

[71] Nogueira-Machado JA, Volpe CMDO, Veloso CA, Chaves MM. HMGB1, TLR and RAGE. Expert Opinion on Therapeutic Targets. 2011;**15**:1023-1035

[72] Čejková S, Králová Lesná I, Poledne R. Monocyte adhesion to the endothelium is an initial stage of atherosclerosis development. Cor et Vasa. 2016;**58**:e419-e425

[73] Wautier JL, Wautier MP. Cellular and Molecular Aspects of Blood Cell-Endothelium Interactions in Vascular Disorders. International Journal of Molecular Sciences. 2020 Jul 27;**21**(15):5315

[74] Sena CM, Matafome P, Louro T, Nunes E, Fernandes R, Seiça RM.

Metformin restores endothelial function in aorta of diabetic rats. British Journal of Pharmacology. 2011;**163**:424-437

[75] Lenna S, Han R, Trojanowska M. Endoplasmic reticulum stress and endothelial dysfunction. IUBMB Life. 2014;**66**:530-537

[76] Carreras-Sureda A, Pihán P, Hetz C. Calcium signaling at the endoplasmic reticulum. Cell Calcium. 2018;**70**:24-31

[77] Battson ML, Lee DM, Gentile CL. Endoplasmic reticulum stress and the development of endothelial dysfunction, American journal of physiology. Heart and Circulatory Physiology. 2017;**312**:H355-H367

[78] Guo Y, Di Shen Y, Zhou Y, Yang J, Liang Y, Zhou N, et al. Deep Learning-Based Morphological Classification of Endoplasmic Reticulum Under Stress. Frontiers in Cell and Developmental Biology. 2021;**9**:767866

[79] Schuck S, Gallagher CM, Walter P. ER-phagy mediates selective degradation of endoplasmic reticulum independently of the core autophagy machinery. Journal of Cell Science. 2014;**127**:4078-4088

[80] Rovira-Llopis S, Bañuls C, Diaz-Morales N, Hernandez-Mijares A, Rocha M, Victor VM. Mitochondrial dynamics in type 2 diabetes. Redox Biology. 2017;**11**:637-645

[81] Park G, Lee JY, Han HM, An HS, Jin Z, Jeong EA, et al. Ablation of dynamin-related protein 1 promotes diabetes-induced synaptic injury in the hippocampus. Cell Death & Disease. 2021;**12**:445

[82] Amodio G, Moltedo O, Faraonio R, Remondelli P. Targeting the endoplasmic reticulum unfolded protein response to counteract the oxidative stress-induced endothelial dysfunction. Oxidative Medicine and Cellular Longevity. 2018;**2018**:4946289

[83] Leem J, Koh EH. Interaction between mitochondria and the endoplasmic reticulum. Experimental Diabetes Research. 2012;**2012**:242984

[84] Rocha M, Apostolova N, Diaz-Rua R, Muntane J, Victor VM. Mitochondria and T2D. Trends in Endocrinology and Metabolism: TEM. 2020;**31**:725-741

[85] Pathomthongtaweechai N, Chutipongtanate S. AGE/RAGE signaling-mediated endoplasmic reticulum stress and future prospects in non-coding RNA therapeutics for diabetic nephropathy. Biomedicine & Pharmacotherapy = Biomedecine & Pharmacotherapie. 2020;**131**:110655

[86] Yang S, Wu M, Li X, Zhao R, Zhao Y, Liu L, et al. Role of endoplasmic reticulum stress in atherosclerosis and its potential as a therapeutic target. Oxidative Medicine and Cellular Longevity. 2020;**2020**:9270107

[87] Wu L, Wang D, Xiao Y, Zhou X, Wang L, Chen B, et al. Endoplasmic reticulum stress plays a role in the advanced glycation end product-induced inflammatory response in endothelial cells. Life Sciences. 2014;**110**:44-51

[88] Xu J, Chen L-J, Yu J, Wang H-J, Zhang F, Liu Q, et al. Involvement of advanced glycation end products in the pathogenesis of diabetic retinopathy. Cellular Physiology and Biochemistry International Journal of Experimental Cellular Physiology, Biochemistry, and Pharmacology. 2018;**48**:705-717

[89] Lan K-C, Chiu C-Y, Kao C-W, Huang K-H, Wang C-C, Huang K-T, et al. Advanced glycation end-products induce apoptosis in pancreatic islet

endothelial cells via NF-κB-activated cyclooxygenase-2/prostaglandin E2 up-regulation. PLoS One. 2015;**10**:e0124418

[90] Zhao G, Zhang X, Wang H, Chen Z. Beta carotene protects H9c2 cardiomyocytes from advanced glycation end product-induced endoplasmic reticulum stress, apoptosis, and autophagy via the PI3K/Akt/mTOR signaling pathway. Annals of Translational Medicine. 2020;**8**:647

[91] Pei Z, Deng Q, Babcock SA, He EY, Ren J, Zhang Y. Inhibition of advanced glycation endproduct (AGE) rescues against streptozotocin-induced diabetic cardiomyopathy. Toxicology Letters. 2018;**284**:10-20

[92] Adegbola P, Aderibigbe I, Hammed W, Omotayo T. Antioxidant and anti-inflammatory medicinal plants have potential role in the treatment of cardiovascular disease. American Journal of Cardiovascular Disease. 2017;**7**:19-32

[93] Kong M, Xie K, Lv M, Li J, Yao J, Yan K, et al. Anti-inflammatory phytochemicals for the treatment of diabetes and its complications. Biomedicine & Pharmacotherapy = Biomedecine & Pharmacotherapie. 2021;**133**:110975

[94] Manco G, Porzio E, Carusone TM. Human Paraoxonase-2 (PON2): Protein Functions and Modulation. Antioxidants. Basel. 2021 Feb 7;**10**(2):256

[95] Yu W, Liu X, Feng L, Yang H, Yu W, Feng T, et al. Glycation of paraoxonase 1 by high glucose instigates endoplasmic reticulum stress to induce endothelial dysfunction in vivo. Scientific Reports. 2017;**7**:45827

[96] Mastorikou M, Mackness B, Liu Y, Mackness M. Glycation of paraoxonase-1 inhibits its activity and impairs the ability of high-density lipoprotein to metabolize membrane lipid hydroperoxides. Diabetic Medicine a Journal of the British Diabetic Association. 2008;**25**:1049-1055

[97] Parween F, Gupta RD. Insights into the role of paraoxonase 2 in human pathophysiology. Journal of Biosciences. 2022;**47**:166

[98] Devarajan A, Grijalva VR, Bourquard N, Meriwether D, Imaizumi S, Shin B-C, et al. Macrophage paraoxonase 2 regulates calcium homeostasis and cell survival under endoplasmic reticulum stress conditions and is sufficient to prevent the development of aggravated atherosclerosis in paraoxonase 2 deficiency/apoE−/− mice on a Western diet. Molecular Genetics and Metabolism. 2012;**107**:416-427

[99] Ravi R, Ragavachetty Nagaraj N, B. Subramaniam Rajesh, effect of advanced glycation end product on paraoxonase 2 expression. Life Sciences. 2020;**246**:117397

[100] Ravi R, Subramaniam Rajesh B. Paraoxonase 2 protects against the CML mediated mitochondrial dysfunction through modulating JNK pathway in human retinal cells, Biochimica et biophysica acta. General Subjects. 2022;**1866**:130043

Chapter 7

# Endoplasmic Reticulum Involvement in Heart Injury: An Overview

*Gaia Favero, Francesca Bonomini, Marzia Gianò and Rita Rezzani*

## Abstract

The Endoplasmic Reticulum (ER) is a multifunctional organelle present in the cytoplasm of the eukaryotic cells. It is involved in many aspects of cellular physiology and it presents important interaction with other cellular organelles. Different physiological and/or pathological factors may alter ER morphology and homeostasis, resulting in the accumulation of a large number of unfolded/misfolded proteins in the ER lumen and so inducing ER stress. Alterations in ER have been found to be related to different disorders. In particular, ER stress is implicated in the development and progression of various heart injuries, such as myocardial infarction, ischemia/reperfusion, heart failure, diabetic cardiomyopathy, arrhythmias and cardiotoxicity. Furthermore, the efficiency to counteract the ER stress declines significantly during the physiopathological aging process. In this chapter, we present the correlation between the ER and cardiac injury focusing mainly on the aging process and then we report a brief overview of the potential involvement of some bioactive molecules as preventive/therapeutic compounds that can contrast heart disorders through ER modulation.

**Keywords:** aging, arrhythmias, biomolecules, cardiac hypertrophy, heart, endoplasmic reticulum

## 1. Introduction

The ER is a complex and multifunctional organelle present in the cytosol of the eukaryotic cells. This organelle is involved in many aspects of cellular physiology such as regulation of calcium, synthesis, transport and folding of protein, synthesis and metabolism of steroids and lipids and, not to be underestimated, it has important interaction with other cellular organelles [1, 2]. The ER is composed of a continuous lipid bilayer that encloses a luminal space [3–6]. It is morphologically characterized by a membrane system with two major shaped domains: the nuclear envelope and the peripheral ER which includes rough leaves and smooth tubules [7]. The nuclear envelope consists of ER membrane wrapped around DNA and various nuclear elements and it is organized in two continuous flat cisternae sheets surrounding the nucleus, defined outer and inner nuclear membranes [6, 8]; while the peripheral ER presents cisternae and an interconnected nuclear network of tubules [5, 9–11]. The nuclear envelope pores selectively control the transport of molecules inside and outside the nucleus. The inner and outer nuclear membranes are ample flat sheet-like cisternae stacked over each other and separated by the internuclear membrane space. Historically, ER cisternae have been classified as ribosome bound: "rough ER - RER" or ribosome-free: "smooth ER - SER" [10]. The RER presents many ribosomes associated to the membrane surface and extends from the nuclear envelope to the cell periphery, so defining specialized areas for protein synthesis, folding and degradation. Furthermore, the outer membrane of the nuclear envelope may be considered as part of the RER domain because it is physically continuous with the RER membranes [12]. The SER is composed of irregular and convoluted tubules without associated ribosomes. However, it is continuous with the RER and the majority of proteins in this compartment come from the RER domain [13]. SER includes membranes specialized for drug metabolism and steroid synthesis, as well as tubulovesicular elements forming ER exit sites [14]. Differences in membrane curvature ulteriorly differentiate the RER and SER subdomains [11]. The different ER membranes organization in domains or regions observed among different cell types correlates strictly with their functions: cells specialized to the production, storage and secretion of proteins, such as exocrine cells, are rich in RER; whereas, endocrine cells that synthesize steroid hormones and muscle cells are rich in SER [15, 16]. Furthermore, the previously described distribution of RER and SER is clearly observed in hepatocytes, neurons and endocrine cells; however, in some cells there is not a clear distinction between both domains and so tubules with associated ribosomes are mixed with tubules without associated ribosomes.

As previously depicted, the structure of the ER is complex due to the various distinct domains present within one continuous membrane bilayer. These domains are shaped by interactions with the cytoskeleton, by proteins that stabilize membrane shape and by a homotypic fusion machinery that allows the ER membrane to maintain its morphology. The ER also contains domains that control the interaction with other organelles, such as the Golgi apparatus, endosomes, mitochondria, lipid droplets and peroxisomes [9, 10, 17]. The purpose of differently shaped ER domains is still under evaluation to better clarify also the related functions. **Figure 1** schematically reported the main ER specialized domains, the ER compartmentalization with the nuclear membrane and its interaction with cellular organelles.

Different physiological and/or pathological factors may alter ER morphology and homeostasis, resulting in the accumulation of a large number of unfolded or misfolded proteins in the ER lumen and so inducing ER stress (**Figure 2**) [18–20]. Alteration in ER has been found to be related with different disorders [4, 21]; in fact, in various pathologies the morphology/structure of ER was abnormal and various mutation in morphology-regulating proteins have been observed, indicating that morphology and functions of the ER are intrinsically linked. Furthermore, in chronic conditions, a persistent ER stress can induce and exacerbate cellular and tissue senescence accelerating the aging-related process [21, 22]. To counteract ER stress and reduce the synthesis of proteins, cells activate an endogenous adaptive stress response defined Unfolded Protein Reaction (UPR). During this adaptive mechanism, ER showed an increased expression of folding protein and degradation of protein related genes [19, 23]. Therefore, ER stress is an adaptive cell mechanism

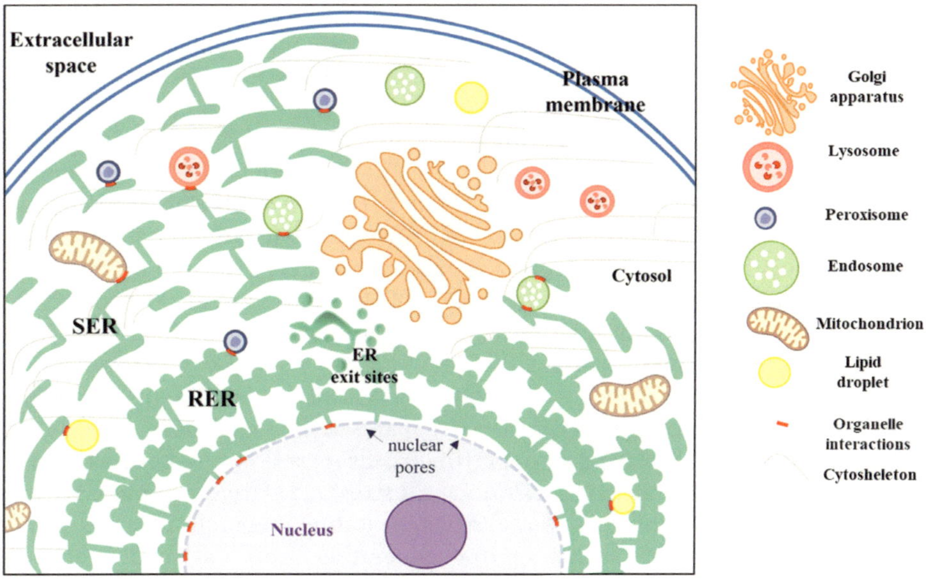


**Figure 1.**
*Schematic representation of the main endoplasmic reticulum specialized domains, the endoplasmic reticulum compartmentalization with the nuclear membrane and its interaction with cellular organelles. RER: Ribosome-free endoplasmic reticulum; SER: Smooth endoplasmic reticulum.*

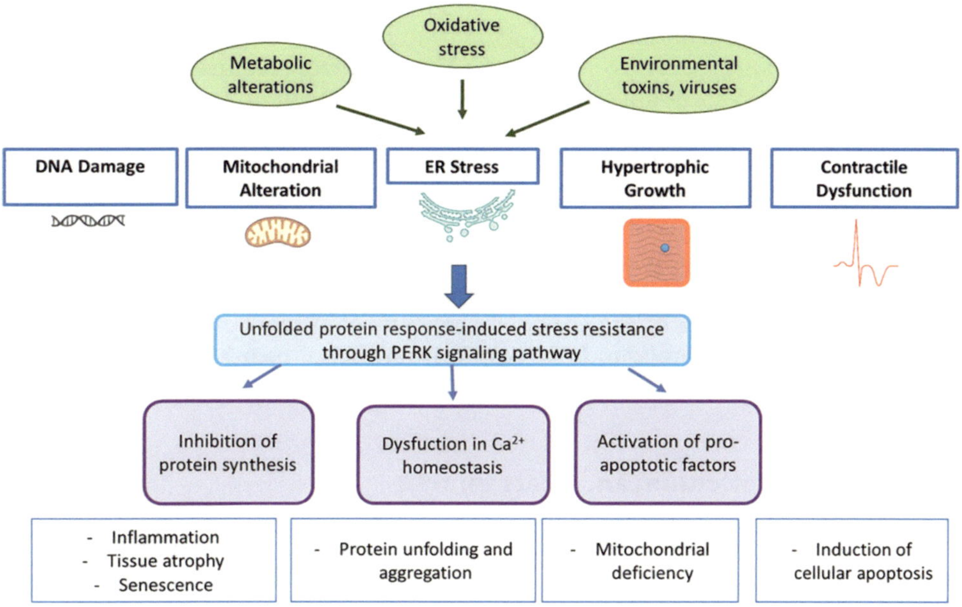


**Figure 2.**
*Cardiomyocyte senescence's markers and regulation of aging process by endoplasmic reticulum stress. Metabolic alterations, oxidative stress, toxins and viruses can induce endoplasmic reticulum stress and activate the unfolded protein reaction via PERK signaling pathway leading to heart injury. ER: Endoplasmic reticulum; PERK: Protein kinase RNA-activated (PKR)-like ER kinase; UPR: Unfolded protein reaction.*

activated to restore the ER morphology and functions by triggering UPR; whereas the persistence of the ER stress is deleterious and induces cell apoptosis [24–26]. To date many studies are focused on ER (stress) as a possible target strategy to counteract various diseases progression and drug resistance by hitting UPR components and the complex interplay between UPR, autophagy and apoptosis. The UPR may be activated by different sensors: Inositol Requiring Enzyme1 (IRE1), Activating Transcription Factor6 (ATF6) and Protein Kinase RNA-activated (PKR)-like ER Kinase (PERK), leading to transcriptional and translational reprogramming to counteract unfolded proteins accumulation and upregulating the expression of UPR-related proteins and proteins of the autophagy-related pathway [27]. PERK is involved in a pro-survival pathway, but it can also promote apoptosis through the activation of CHOP, the main mediator of apoptosis induced by the UPR [28, 29]. CHOP induces the expression of ER Oxidoreductase 1 alpha (ERO1α), which promotes cell death through the hyper-oxidation of ER proteins [30]. CHOP also participates in the regulation of autophagy, self-digestive intracellular process through which cells recycle organelles and damaged or unnecessary proteins [31–33]. Prolong ER stress, oxidative stress and inflammation are typical features of many pathologies, including metabolic diseases and different age-related diseases, such as neurodegenerative disorders, various cancers and cardiovascular diseases [22, 34]. Various studies have demonstrated that damage at myocardial tissue level is characterized by accumulation, at ER level, of ubiquitinated proteins and this accumulation at ER level promotes ER stress which, in turn, induces the release of ER-related apoptotic proteins thus contributing, in a vicious cycle, to heart injury [35–37].

Furthermore, autophagy is a crucial process involved also in the development of cardiovascular diseases [38]. A variety of autophagy proteins are localized at the ER level and originate from mitochondrial-associated ER membrane (ER interaction site with mitochondria) [39]. ER is one of the membrane donors required for the formation of autophagosomes, double-membrane vesicles involved in autophagy [40, 41].

Furthermore, ER stress is progressively increased during the physiopathological aging process also at heart level. The modulation of ER stress and an enhancement of the intensity of UPR adaptive stress response could be valid strategies to prevent/treat different diseases.

In the following paragraphs, we present the main ER alterations correlated with cardiac injury focusing mainly on the physiopathological process of aging and then we report a brief overview of the potential involvement of bioactive molecules as preventive/therapeutic compounds that can contrast heart disorders through ER modulation.

## 2. Endoplasmic reticulum and heart injury

Cardiovascular diseases are to date the leading morbidity and mortality burden in the world [42–44]. The molecular mechanism(s) involved in the pathogenetic processes of cardiac disorders are not fully understood. Age-related changes contribute to reduce heart capacity to adapt to different physiological and pathological factors and, in turn, the ability to restore the "normal" morphology and functions is reduced/impaired [45, 46]. Notably, aging promotes alteration of mitochondrial function, condition strictly linked to increased reactive oxygen species production and oxidative stress (which play a fundamental role in aging-related diseases). In addition, elevated levels of oxidative and nitrosative stress occur in parallel with ER stress [45, 47] and, therefore, aging-induced ER stress may contribute to mitochondrial dysfunction [36]. In the last decade, various studies have reported that ER stress, mitochondrial injury and oxidative stress are involved in the development and progression of a various heart diseases, such as myocardial infarction, cardiac hypertrophy, ischemia/reperfusion, heart failure, diabetic cardiomyopathies, arrhythmias and cardiotoxicity [48–51]. Furthermore, the cell ability to counteract the ER stress and the efficacy of the UPR signaling adaptive response decline significantly during aging [45]. The UPR is, in fact, fundamental for maintaining heart health because cardiomyocytes lack the ability to replicate or efficiently repair themselves, making them deficient in regenerative capability [52]. If ER stress becomes too severe, the UPR signaling adaptive response leads to cell death through apoptosis or autophagy processes [29, 53]. During ER stress, IRE1 binds to Tumor Necrosis Factor-Receptor-Associated Factor 2 (TRAF2) which in turn activates the Apoptosis Signal-Regulating Kinase 1 (ASK1), a Mitogen-Activated Protein (MAP) Kinase Kinase Kinase (MAP3K). The IRE1-TRAF2-ASK1 complex promotes the activation of p38 and c-Jun N-terminal Kinase (JNK), known apoptosis-regulating MAP kinases (MAPK) [54]. In particular, JNK and p38 may be activated in response to a variety of cell stresses, including oxidative stress, DNA damage and inflammatory cytokines [55]. Under physiological conditions, glucose-regulated protein (GRP)78, a mediator of the UPR, binds to PERK, IRE1 or ATF6 forming a stable and inactive complex which inhibits the transmission of the downstream signals. When the UPR is promoted, GRP78 disassociates from this complex and binds to unfolded or misfolded proteins, activating downstream cascade reactions and promoting cell death [56–58]. It is fundamental to underline that this adaptive process involves a complex correlation between autophagy and apoptosis, fundamental to determine cell fate in response to ER stress. In fact, autophagy promotes cell survival through degradation of misfolded proteins that have been retained in the ER or induces cell death through the autophagy process [52]. Understanding how to suppress IRE1 activity and related downstream pathway to inhibit ER stress-induced apoptosis and autophagy-related cell death is of fundamental importance. Xue et al. [59] observed that Activin A, a member of the transforming growth factor-beta superfamily, has important neuroprotective effects against ER stress-mediated apoptosis and autophagy by inhibiting the activation of the IRE1-TRAF2-ASK1-JNK/p38 complex *in vitro*. This will be an important starting point for better understanding how to modulate ER stress and related injuries.

Therefore, investigate the ER stress also in aging-related heart injury is actually an important research focus which needs more studies.

For the first time, Sreedhar et al. [60] identified the involvement of ER stress-induced apoptosis in the hearts of SAMP8 mice, known aging mouse model [61, 62], demonstrating that ER is an important factor in cardiac aging. In particular, SAMP8 mice showed an increased p38 expression at heart level, suggesting a relation between MAPK signaling cascade and ER stress. At heart level, the SAMP8 mice presented elevated expression of the pro-apoptotic transcription factor CHOP and caspases, highlighting that p38 induces ER stress-related apoptosis. Currently, a weak/moderate ER stress is considered beneficial for heart function by restoring ER morphology and homeostasis and protecting cardiomyocytes through an adaptive mechanism. On the contrary, a prolonged and severe ER stress promotes cardiomyocyte death [63–65]. Notably, mitochondrial damage and ER stress have been well recognized as important cardiac ischemia/reperfusion injury upstream factors controlling the cardiomyocyte death [20, 66]. Aging decreases the efficacy of the endogenous adaptive response and increases ER-mediated myocardial apoptotic signaling after ischemic/reperfusion injuries [24, 67].

Cardiomyopathy is the most common cause of morbidity and mortality in diabetic patients [68]. Diabetic conditions promote ER stress response by oxidative mechanisms which contribute to eliminate unhealthy cells and lead to diabetic cardiomyopathy development. Such conditions have been shown to promote cardiac hypertrophy, collagen deposition, stiffness and so important cardiac dysfunction. ER stress activation indeed has been observed in both hypertrophic and failing hearts [20, 68, 69]. Accumulating studies have demonstrated that the ER stress response is also involved in the cardiac hypertrophy progression which is typically characterized by increased size of cardiomyocyte, interstitial fibrosis, apoptosis and contractile dysfunction [43, 70]. STING, an ER-resident protein regulating innate immunity, is highly expressed in hearts of patients with dilated cardiomyopathy or hypertrophic cardiomyopathy [71]. In addition, Zhang et al. [71] reported that STING knockout attenuated cardiac hypertrophy induced by aortic banding showing a significant alteration of cardiomyocyte size, ER stress and hypertrophic markers expression. Moreover, STING deletion significantly reduced inflammation and fibrosis at heart level. The authors so concluded that the modulation of STING expression was, at least in part, regulated by ER stress, underling its important contribution in the cardiac hypertrophy progression. Furthermore, Yao et al. reported a fundamental physiological role of AGGF1: it may regulate ER stress signaling and blocking ER stress-induced apoptosis in cardiac hypertrophy [51]. The authors described that ER stress induces the downregulation of AGGF1 level in a mouse model and human patients with heart failure. In detail, AGGF1 regulates ER stress signaling through CHOP pathway and so inhibits ER stress-induced apoptosis. Liu et al. [72] observed that the expression of SOCS3, a mechanical stress inducible gene, is involved in hypertrophic hearts after 2 weeks of transverse aortic constriction, well-established animal model of pressure overload-induced cardiac hypertrophy [72, 73]. Pressure overload promotes ER stress-induced apoptosis of cardiomyocytes, leading to cardiac hypertrophy and heart failure [74]. In detail, the authors observed that SOCS3-GRP78-ER stress signaling promotes the transition from cardiac hypertrophy to heart failure during pressure overload *in vivo*. In fact, prolonged pressure overload significantly decreases SOCS3 expression that, acting as a negative regulator of cardiac hypertrophy, interacts with GRP78 and induces GRP78 ubiquitination and proteasomal degradation so modulating ER stress. Moreover, cardiac-specific SOCS3 knockout mice presented significant cardiac hypertrophy, chamber dilatation and abnormal myofilament calcium sensitivity after pressure overload. Notably, the treatment of mice with 4-phenylbutyric acid, a short chain fatty acid that is clinically used to treat urea cycle disorders, for 4 weeks attenuates ER stress and related downstream pathways targeting GRP78 and so determining inhibition of cardiac hypertrophy and dysfunction.

STING, AGGF1 and SOCS3 may represent new targets not only for cardiovascular diseases, but also for other diseases associated with ER stress.

During aging, the proteostasis network becomes unable to maintain proteostasis and key UPR molecules, such as PERK, are damaged leading to misfolded proteins accumulation within the ER [75–78]. Age-related injury due to protein misfolding, aggregating proteins and dysfunctional UPR sensors has been shown to lead to diabetes mellitus, neurodegeneration, cancer, heart diseases and arrhythmias [75, 77]. Due to the strict morphological and functional link between ER protein and calcium homeostasis, mounting evidences report that prolonged ER stress is correlated with heart arrhythmic risk via perturbed redox status within the ER as well as downregulation of cardiac ion channel proteins. Furthermore, UPR is involved directly and indirectly in proarrhythmic cardiac remodeling through UPR-induced oxidative stress, altered glycosylation and modulation of ion channels involved in excitation-contraction (including ER calcium handling proteins) [72], so ulteriorly correlating ER stress to heart arrhythmias. Importantly, inhibition of PERK may prevent downregulation of these channels, attenuate aberrant electrical remodeling, reduce ventricular arrhythmia inducibility and improve survival after myocardial infarct *in vivo*. Therefore, it is fundamental that cardiomyocytes carefully balance the UPR to safely counteract ER stress and maintain proteostasis [78]. Recently, Nakamura et al. [70] reported that ER stress, activating nuclear factor-kappa B (NF-kB) pathway, promotes ventricular arrhythmia in failing hearts via the cardiac dopamine receptor D1 (D1R) upregulation. D1R is upregulated in cardiomyocytes of failing hearts (in mice and humans) leading to heart failure-associated ventricular arrhythmia. ER stress-induced NF-κB signaling pathway could potentially serve as a target to improve the prognosis of heart failure patients [79]. Moreover, Hamilton et al. [80] recently observed a novel association interacting complex Ryanodine Receptor (RyR2)-Endoplasmic Reticulum protein44 (ERp44) which may be stabilized by Ero1α in cardiac hypertrophy. Cardiac hypertrophy-mediated upregulation of Ero1α results in the removal of ERp44 from the complex, inducing RyR2 dysfunction. This dysfunction increases the risk of calcium-dependent ventricular arrhythmias. The authors, notably, concluded that Ero1α may be a promising target to reduce arrhythmogenesis and to improve cardiac function during hypertrophy and heart failure, without disturbing the finely balanced intra-ER redox environment.

Autophagy is also impaired in arrhythmogenic cardiomyopathy. Pitsch et al. [81] observed, in an animal model of arrhythmogenic cardiomyopathy, signs of increased autophagy prior to structural disease onset in the "normal"-appearing mutant myocardium. In particular, the authors reported, at heart level, numerous autophagy-related vacuoles and the upregulation of autophagy during onset and progression of the arrhythmogenic cardiomyopathy. Furthermore, ventricles presented elevated expression of the pro-apoptotic ER stress marker CHOP both at disease onset and during chronic disease progression. In addition, reduced Ryr2 mRNA expression together with severe enlarged ER cisternae underlined ER dysfunction.

Furthermore, it has been documented that ER stress is involved also in cardiotoxicity and heart injury that potentially may evolve into heart failure. Interestingly, Ni et al. [82] observed upregulation of ER chaperones in patients with end-stage heart failure. In detail, the authors investigated the expression of various ER stress factors in hearts of 4 patients subjected to heart transplantation who suffered from dilated cardiomyopathy with end-stage heart failure and 9 patients undergoing mitral valve replacement, as well as 4 healthy subjects. This study, interestingly, reported a significant upregulation of the phosphorylated level of PERK together with c-Jun phosphorylation in failing hearts compared with healthy hearts. These results indicated that prolonged ER stress and associated apoptosis are general occurrence in human failing hearts. Similar results were subsequently found in isoproterenol-stimulated cardiomyocytes and in a rat model of heart failure after abdominal aortic constriction and isoproterenol subcutaneous injection. Furthermore, the authors observed also that long-term oral treatment with β-adrenergic receptor blockers inhibits ER stress, correlated to cardiac hypertrophy and heart failure.

## 3. Possible endoplasmic reticulum modulation against heart injury

To date, the protective and therapeutic strategies against cardiovascular diseases are not fully efficient and effective especially in elderly people, so amplify the actual knowledge on ER stress involvement in heart injury will be helpful in finding novel drug targets. In the last decade, at this aim, different potential therapeutic biomolecules have been investigated; however, there are still many unresolved questions that need to be debated and investigated in deep.

Different studies have indicated that sirtuin 1, $NAD^+$-dependent deacetylases implicated in many aspects of aging process [83], is involved in the regulation of ER stress in cardiomyocytes [84–86]. Hsu et al. [45] hypothesized that sirtuin 1 is able to protect aging heart. In detail, the authors observed a significative contractile dysfunction associated with improved ER stress and oxidative stress in aged sirtuin $1^{-/-}$ mouse hearts. The study demonstrated also, *in vitro*, that a sirtuin 1 activator reduced ER stress and myocardial apoptosis induced by oxidative stress; whereas a sirtuin 1 inhibitor reversed the sirtuin 1 protective effect at cardiomyocytes level. Therefore, the development of "drugs" targeting sirtuin 1 could have interesting preventive/therapeutic potentials against cardiac contractile injury aging-related.

Recently, Monceaux et al. [63] evaluated the ability of ten phenolic phytonutrients (resveratrol, berberine, butein, catechin, ferulic acid, isoliquiritigenin, malvidin, piceatannol, pterostilbene and tyrosol) to modulate ER stress. Interestingly, all these bioactive compounds are able to protect the heart from severe ER stress. Investigating in deep the mechanisms of action of the phytonutrients, the authors reported that ferulic acid, pterostilbene and tyrosol protect cardiomyocytes from severe ER stress by selectively downregulating the PERK pathway of the UPR signaling through sirtuin 1-mediated deacetylation of the translation initiation factor eIF2α (factor which modulates the transcription of UPR target genes, such as ATF4). Interestingly, this study reported that ferulic acid, pterostilbene and tyrosol, by downregulating the PERK/eIF2α/ATF4 pathway, reduce the level of the pro-apoptotic transcription factor CHOP and so limit ER stress-related apoptosis at heart level (**Figure 3**).

Furthermore, Liu et al. [72] evaluated the protective role in reducing the activation of the UPR under ER stress conditions of Wenxin Granules (mix of five drugs: Codonopsis, Rhizoma Polygoni, Panax notoginseng, Amber and Gansong), known traditional Chinese medicine with important effects in the inhibition of myocardial remodeling, modulation of cardiac conduction systems and heart arrhythmias [87, 88]. The authors performed an in vivo study with a rat model of myocardial infarction obtained through the ligation of the anterior descending branch of the left coronary artery and Wenxin Granules were administered intragastrically once per day for two consecutive weeks. In detail, the human daily dose of Wenxin Granules was converted into an equivalent dose for rats (approximately 6 times the human dose for the low-dose group and approximately 12 times the human dose for the high-dose group). Compared with the sham operated group, the myocardial infarction model group showed larger hearts and a significant increase of the left ventricular inner diameter; whereas Wenxin Granules improved significantly the morphological myocardial alteration observed in the infarction model group. Furthermore, the myocardial infarction model group presented elevated expression levels of the ER stress proteins GRP78, PERK, ATF6 and XBP1 and, notably, the expression levels of the ER stress pathway proteins were reduced after Wenxin Granules administration. Furthermore, the expression levels of the apoptotic CHOP and Bax in the myocardial infarction model group significantly increased compared with the sham group, whereas the Bcl-2/Bax ratio significantly decreased. Notably, Wenxin Granules were able to improve ventricular remodeling, prevented the excessive ER stress-mediated UPR activation and inhibited myocardial cell apoptosis [72].

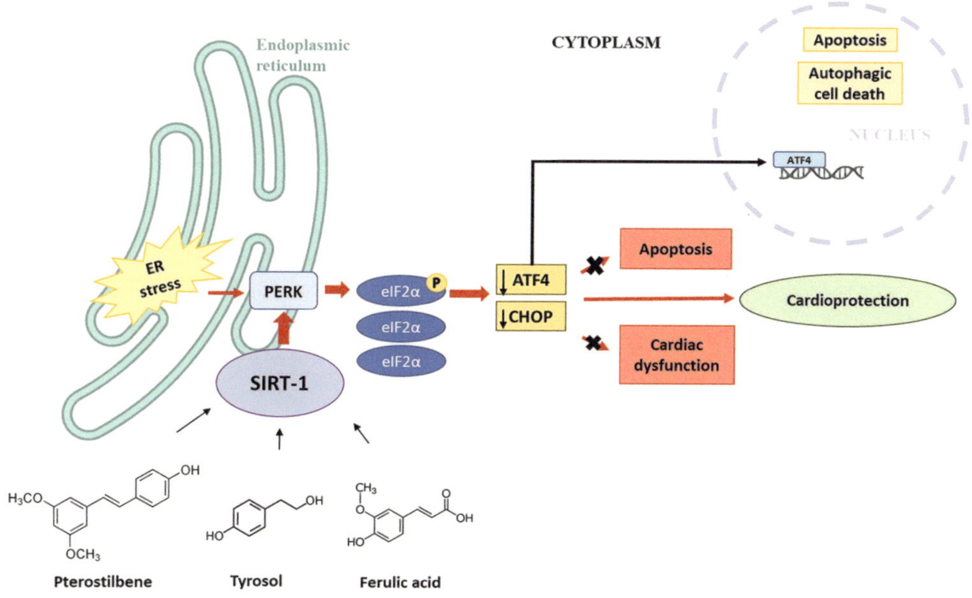


**Figure 3.**
*Polyphenol cardioprotection's mechanism against endoplasmic reticulum stress. Pterostilbene, tyrosol and ferulic acid protect cardiomyocytes downregulating the PERK signaling pathway mediated by sirtuin 1. This pathway decreases the expression of ATF4 and CHOP inhibiting apoptosis and cardiac dysfunction. Modified from Monceaux et al. [63]. ATF4: Activating transcription factor4; ER: Endoplasmic reticulum; PERK: Protein kinase RNA-activated (PKR)-like ER kinase.*

Recently, Tu et al. [89] analyzed for the first time the effects of fish oil against atrial fibrillation vulnerability by reducing ER stress in a canine model. The atrial fibrillation model animals were obtained using long-term rapid atrial pacing; the animals were fed with a chow supplemented with fish oil (0.6 dietary ω-3 fatty acids/kg/day) and initiated the treatment 1 week before surgery and continued for 4 weeks post-surgery. The fish oil treatment not only reduced myocardial ER stress, but decreased also pro-inflammatory factors, calcium handling-related proteins and reversed the elevated level of CHOP and caspase12 in the atrial fibrillation group, so significantly reduced atrial fibrillation inducibility and duration.

## 4. Conclusion

The actual knowledge on the role of ER stress in heart injury are mainly from *in vivo* animal studies, so the understanding of the ER and UPR involvement in human diseases is still limited. Due to that, ER stress plays a central role in heart injury development, ER stress inhibitors could be important for preventive/therapeutic interventions. The development of novel therapeutic/protective approaches modulating the ER stress pathway may reduce the drain of health care costs and resources resulting from the worldwide spread against cardiovascular diseases.

## Author details

Gaia Favero[1,2*], Francesca Bonomini[1,2,3], Marzia Gianò[1] and Rita Rezzani[1,2,3]

1 Anatomy and Physiopathology Division, Department of Clinical and Experimental Sciences, University of Brescia, Italy

2 Interdepartmental University Center of Research "Adaptation and Regeneration of Tissues and Organs (ARTO)", University of Brescia, Italy

3 Italian Society of Orofacial Pain (SISDO), Brescia, Italy

*Address all correspondence to: gaia.favero@unibs.it

## References

[1] Cohen S, Valm AM, Lippincott-Schwartz J. Interacting organelles. Current Opinion in Cell Biology. 2018;**53**:84-91. DOI: 10.1016/j.ceb.2018.06.003

[2] Schwarz DS, Blower MD. The endoplasmic reticulum: Structure, function and response to cellular signaling. Cellular and Molecular Life Sciences. 2016;**73**:79-94. DOI: 10.1007/s00018-015-2052-6

[3] Bai L, You Q, Feng X, Kovach A, Li H. Structure of the ER membrane complex, a transmembrane-domain insertase. Nature. 2020;**584**:475-478. DOI: 10.1038/s41586-020-2389-3

[4] Perkins HT, Allan V. Intertwined and finely balanced: Endoplasmic reticulum morphology, dynamics, function, and diseases. Cell. 2021;**10**:2341. DOI: 10.3390/cells10092341

[5] Sun S, Tang X, Guo Y, Hu J. Endoplasmic reticulum composition and form: Proteins in and out. Current Opinion in Cell Biology. 2021;**71**:1-6. DOI: 10.1016/j.ceb.2021.01.008

[6] Voeltz GK, Prinz WA, Shibata Y, Rist JM, Rapoport TA. A class of membrane proteins shaping the tubular endoplasmic reticulum. Cell. 2006;**124**:573-586. DOI: 10.1016/j.cell.2005.11.047 Erratum in: Cell. 2007;130(4):754

[7] Silva-Palacios A, Zazueta C, Pedraza-Chaverri J. ER membranes associated with mitochondria: Possible therapeutic targets in heart-associated diseases. Pharmacological Research. 2020;**156**:104758. DOI: 10.1016/j.phrs.2020.104758

[8] Lu L, Ladinsky MS, Kirchhausen T. Cisternal organization of the endoplasmic reticulum during mitosis. Molecular Biology of the Cell. 2009;**20**(15):3471-3480. DOI: 10.1091/mbc.e09-04-0327

[9] Almeida C, Amaral MD. A central role of the endoplasmic reticulum in the cell emerges from its functional contact sites with multiple organelles. Cellular and Molecular Life Sciences. 2020;**77**:4729-4745. DOI: 10.1007/s00018-020-03523-w

[10] English AR, Voeltz GK. Endoplasmic reticulum structure and interconnections with other organelles. Cold Spring Harbor Perspectives in Biology. 2013;**5**:a013227. DOI: 10.1101/cshperspect.a013227

[11] Shibata Y, Voeltz GK, Rapoport TA. Rough sheets and smooth tubules. Cell. 2006;**126**:435-439. DOI: 10.1016/j.cell.2006.07.019

[12] Mehrtash AB, Hochstrasser M. Ubiquitin-dependent protein degradation at the endoplasmic reticulum and nuclear envelope. Seminars in Cell & Developmental Biology. 2019;**93**:111-124. DOI: 10.1016/j.semcdb.2018.09.013

[13] Sanvictores T, Davis DD. Histology, Rough Endoplasmic Reticulum. Treasure Island (FL): StatPearls Publishing; 2023

[14] Balla T, Kim YJ, Alvarez-Prats A, Pemberton J. Lipid dynamics at contact sites between the endoplasmic reticulum and other organelles. Annual Review of Cell and Developmental Biology. 2019;**35**:85-109. DOI: 10.1146/annurev-cellbio-100818-125251

[15] Conte E, Dinoi G, Imbrici P, De Luca A, Liantonio A. Sarcoplasmic

reticulum Ca2+ buffer proteins: A focus on the yet-to-Be-explored role of sarcalumenin in skeletal muscle health and disease. Cell. 2023;**12**(5):715. DOI: 10.3390/cells12050715

[16] Wray S, Burdyga T. Sarcoplasmic reticulum function in smooth muscle. Physiological Reviews. 2010;**90**(1):113-178. DOI: 10.1152/physrev.00018.2008

[17] Lee JE, Cathey PI, Wu H, Parker R, Voeltz GK. Endoplasmic reticulum contact sites regulate the dynamics of membraneless organelles. Science. 2020;**367**:eaay7108. DOI: 10.1126/science.aay7108

[18] Hetz C, Zhang K, Kaufman RJ. Mechanisms, regulation and functions of the unfolded protein response. Nature Reviews. Molecular Cell Biology. 2020;**21**:421-438. DOI: 10.1038/s41580-020-0250-z

[19] Li B, Zhang T, Tang M. Toxicity mechanism of nanomaterials: Focus on endoplasmic reticulum stress. Science of the Total Environment. 2022;**834**:155417. DOI: 10.1016/j.scitotenv.2022.155417

[20] Yarmohammadi F, Hayes AW, Karimi G. The cardioprotective effects of hydrogen sulfide by targeting endoplasmic reticulum stress and the Nrf2 signaling pathway: A review. BioFactors. 2021;**47**:701-712. DOI: 10.1002/biof.1763

[21] Ghemrawi R, Khair M. Endoplasmic reticulum stress and unfolded protein response in neurodegenerative diseases. International Journal of Molecular Sciences. 2020;**21**:6127. DOI: 10.3390/ijms21176127

[22] Salminen A, Kaarniranta K, Kauppinen A. ER stress activates immunosuppressive network: Implications for aging and Alzheimer's disease. Journal of Molecular Medicine (Berlin, Germany). 2020;**98**:633-650. DOI: 10.1007/s00109-020-01904-z

[23] Read A, Schröder M. The unfolded protein response: An overview. Biology (Basel). 2021;**10**:384. DOI: 10.3390/biology10050384

[24] Demirel-Yalciner T, Sozen E, Ozer NK. Endoplasmic reticulum stress and miRNA impairment in aging and age-related diseases. Frontiers in Aging. 2022;**2**:790702. DOI: 10.3389/fragi.2021.790702

[25] Hossain MS, Kader MA, Goh KW, Islam M, Khan MS, Harun-Ar Rashid M, et al. Herb and spices in colorectal cancer prevention and treatment: A narrative review. Frontiers in Pharmacology. 2022;**13**:865801. DOI: 10.3389/fphar.2022.865801

[26] Rellmann Y, Eidhof E, Dreier R. Review: ER stress-induced cell death in osteoarthritic cartilage. Cellular Signalling. 2021;**78**:109880. DOI: 10.1016/j.cellsig.2020.109880

[27] Pires Da Silva J, Monceaux K, Guilbert A, Gressette M, Piquereau J, Novotova M, et al. SIRT1 protects the heart from ER stress-induced injury by promoting eEF2K/eEF2-dependent autophagy. Cell. 2020;**9**(2):426. DOI: 10.3390/cells9020426

[28] Sano R, Reed JC. ER stress-induced cell death mechanisms. Biochimica et Biophysica Acta. 2013;**1833**(12):3460-3470. DOI: 10.1016/j.bbamcr.2013.06.028

[29] Szegezdi E, Logue SE, Gorman AM, Samali A. Mediators of endoplasmic reticulum stress-induced apoptosis. EMBO Reports. 2006;**7**(9):880-885. DOI: 10.1038/sj.embor.7400779

[30] Sisinni L, Pietrafesa M, Lepore S, Maddalena F, Condelli V, Esposito F, et

al. Endoplasmic reticulum stress and unfolded protein response in breast cancer: The balance between apoptosis and autophagy and its role in drug resistance. International Journal of Molecular Sciences. 2019;**20**(4):857. DOI: 10.3390/ijms20040857

[31] Bahamondes Lorca VA, Wu S. Ultraviolet light, unfolded protein response and autophagy. Photochemistry and Photobiology. 2023;**99**(2):498-508. DOI: 10.1111/php.13777

[32] Preston GM, Brodsky JL. The evolving role of ubiquitin modification in endoplasmic reticulum-associated degradation. The Biochemical Journal. 2017;**474**(4):445-469. DOI: 10.1042/BCJ20160582

[33] Wek RC, Cavener DR. Translational control and the unfolded protein response. Antioxidants & Redox Signaling. 2007;**9**(12):2357-2371. DOI: 10.1089/ars.2007.1764

[34] Xiang C, Wang Y, Zhang H, Han F. The role of endoplasmic reticulum stress in neurodegenerative disease. Apoptosis. 2017;**22**:1-26. DOI: 10.1007/s10495-016-1296-4

[35] Jong CJ, Ito T, Prentice H, Wu JY, Schaffer SW. Role of mitochondria and endoplasmic reticulum in taurine-deficiency-mediated apoptosis. Nutrients. 2017;**9**:795. DOI: 10.3390/nu9080795

[36] Thompson J, Maceyka M, Chen Q. Targeting ER stress and calpain activation to reverse age-dependent mitochondrial damage in the heart. Mechanisms of Ageing and Development. 2020;**192**:111380. DOI: 10.1016/j.mad.2020.111380

[37] Wang Y, Cui L, Xu H, Liu S, Zhu F, Yan F, et al. TRPV1 agonism inhibits endothelial cell inflammation via activation of eNOS/NO pathway. Atherosclerosis. 2017;**260**:13-19. DOI: 10.1016/j.atherosclerosis.2017.03.016

[38] Del Re DP, Amgalan D, Linkermann A, Liu Q, Kitsis RN. Fundamental mechanisms of regulated cell death and implications for heart disease. Physiological Reviews. 2019;**99**(4):1765-1817. DOI: 10.1152/physrev.00022.2018

[39] Qu J, Li M, Li D, Xin Y, Li J, Lei S, et al. Stimulation of Sigma-1 receptor protects against cardiac fibrosis by alleviating ire1 pathway and autophagy impairment. Oxidative Medicine and Cellular Longevity. 2021;**2021**:8836818. DOI: 10.1155/2021/8836818

[40] Hamasaki M, Furuta N, Matsuda A, Nezu A, Yamamoto A, Fujita N, et al. Autophagosomes form at ER-mitochondria contact sites. Nature. 2013;**495**(7441):389-393. DOI: 10.1038/nature11910

[41] Wei Y, Liu M, Li X, Liu J, Li H. Origin of the autophagosome membrane in mammals. BioMed Research International. 2018;**2018**:1-9. DOI: 10.1155/2018/1012789

[42] Andersson C, Vasan RS. Epidemiology of cardiovascular disease in young individuals. Nature Reviews. Cardiology. 2018;**15**:230-240. DOI: 10.1038/nrcardio.2017.154

[43] Liu S, Li Y, Zeng X, Wang H, Yin P, Wang L, et al. Burden of cardiovascular diseases in China, 1990-2016: Findings from the 2016 global burden of disease study. JAMA Cardiology. 2019;**4**:342-352. DOI: 10.1001/jamacardio.2019.0295

[44] Zhao D, Liu J, Wang M, Zhang X, Zhou M. Epidemiology of cardiovascular disease in China: Current features

and implications. Nature Reviews. Cardiology. 2019;**16**:203-212. DOI: 10.1038/s41569-018-0119-4

[45] Hsu YJ, Hsu SC, Hsu CP, Chen YH, Chang YL, Sadoshima J, et al. Sirtuin 1 protects the aging heart from contractile dysfunction mediated through the inhibition of endoplasmic reticulum stress-mediated apoptosis in cardiac-specific Sirtuin 1 knockout mouse model. International Journal of Cardiology. 2017;**228**:543-552. DOI: 10.1016/j.ijcard.2016.11.247

[46] Lakatta EG, Sollott SJ. Perspectives on mammalian cardiovascular aging: Humans to molecules. Comparative Biochemistry and Physiology. Part A, Molecular & Integrative Physiology. 2002;**132**:699-721. DOI: 10.1016/s1095-6433(02)00124-1

[47] Pagan LU, Gomes MJ, Gatto M, Mota GAF, Okoshi K, Okoshi MP. The role of oxidative stress in the aging heart. Antioxidants (Basel). 2022;**11**:336. DOI: 10.3390/antiox11020336

[48] Mao W, Fukuoka S, Iwai C, Liu J, Sharma VK, Sheu SS, et al. Cardiomyocyte apoptosis in autoimmune cardiomyopathy: Mediated via endoplasmic reticulum stress and exaggerated by norepinephrine. American Journal of Physiology. Heart and Circulatory Physiology. 2007;**293**:H1636-H1645. DOI: 10.1152/ajpheart.01377.2006

[49] Minamino T, Kitakaze M. ER stress in cardiovascular disease. Journal of Molecular and Cellular Cardiology. 2010;**48**:1105-1110. DOI: 10.1016/j.yjmcc.2009.10.026

[50] Ochoa CD, Wu RF, Terada LS. ROS signaling and ER stress in cardiovascular disease. Molecular Aspects of Medicine. 2018;**63**:18-29. DOI: 10.1016/j.mam.2018.03.002

[51] Yao Y, Lu Q, Hu Z, Yu Y, Chen Q, Wang QK. A non-canonical pathway regulates ER stress signaling and blocks ER stress-induced apoptosis and heart failure. Nature Communications. 2017;**8**:133. DOI: 10.1038/s41467-017-00171-w

[52] Ren J, Bi Y, Sowers JR, Hetz C, Zhang Y. Endoplasmic reticulum stress and unfolded protein response in cardiovascular diseases. Nature Reviews. Cardiology. 2021;**18**:499-521. DOI: 10.1038/s41569-021-00511-w

[53] Momoi T. Conformational diseases and ER stress-mediated cell death: Apoptotic cell death and autophagic cell death. Current Molecular Medicine. 2006;**6**(1):111-118. DOI: 10.2174/156652406775574596

[54] Xin Q, Ji B, Cheng B, Wang C, Liu H, Chen X, et al. Endoplasmic reticulum stress in cerebral ischemia. Neurochemistry International. 2014;**68**:18-27. DOI: 10.1016/j.neuint.2014.02.001

[55] Nozaki K, Nishimura M, Hashimoto N. Mitogen-activated protein kinases and cerebral ischemia. Molecular Neurobiology. 2001;**23**(1):1-19. DOI: 10.1385/MN:23:1:01

[56] Arrieta A, Blackwood EA, Glembotski CC. ER protein quality control and the unfolded protein response in the heart. Current Topics in Microbiology and Immunology. 2018;**414**:193-213. DOI: 10.1007/82_2017_54

[57] Choy KW, Murugan D, Mustafa MR. Natural products targeting ER stress pathway for the treatment of cardiovascular diseases. Pharmacological Research. 2018;**132**:119-129. DOI: 10.1016/j.phrs.2018.04.013

[58] Wang X, Xu L, Gillette TG, Jiang X, Wang ZV. The unfolded protein response in ischemic heart disease. Journal of Molecular and Cellular Cardiology. 2018;**117**:19-25. DOI: 10.1016/j.yjmcc.2018.02.013

[59] Xue LX, Liu HY, Cui Y, Dong Y, Wang JQ, Ji QY, et al. Neuroprotective effects of Activin a on endoplasmic reticulum stress-mediated apoptotic and autophagic PC12 cell death. Neural Regeneration Research. 2017;**12**(5):779-786. DOI: 10.4103/1673-5374.206649

[60] Sreedhar R, Giridharan VV, Arumugam S, Karuppagounder V, Palaniyandi SS, Krishnamurthy P, et al. Role of MAPK-mediated endoplasmic reticulum stress signaling in the heart during aging in senescence-accelerated prone mice. BioFactors. 2016;**42**:368-375. DOI: 10.1002/biof.1280

[61] Davis JA, Paul JR, Mokashi MV, Yates SA, Mount DJ, Munir HA, et al. Circadian disruption of hippocampus in an early senescence male mouse model. Pharmacology, Biochemistry, and Behavior. 2022;**217**:173388. DOI: 10.1016/j.pbb.2022.173388

[62] Lanz M, Janeiro MH, Milagro FI, Puerta E, Ludwig IA, Pineda-Lucena A, et al. Trimethylamine N-oxide (TMAO) drives insulin resistance and cognitive deficiencies in a senescence accelerated mouse model. Mechanisms of Ageing and Development. 2022;**204**:111668. DOI: 10.1016/j.mad.2022.111668

[63] Monceaux K, Gressette M, Karoui A, Pires Da Silva J, Piquereau J, Ventura-Clapier R, et al. Ferulic acid, pterostilbene, and tyrosol protect the heart from ER-stress-induced injury by activating SIRT1-dependent deacetylation of eIF2α. International Journal of Molecular Sciences. 2022;**23**:6628. DOI: 10.3390/ijms23126628

[64] Prola A, Nichtova Z, Pires Da Silva J, Piquereau J, Monceaux K, Guilbert A, et al. Endoplasmic reticulum stress induces cardiac dysfunction through architectural modifications and alteration of mitochondrial function in cardiomyocytes. Cardiovascular Research. 2019;**115**:328-342. DOI: 10.1093/cvr/cvy197

[65] Wang S, Binder P, Fang Q, Wang Z, Xiao W, Liu W, et al. Endoplasmic reticulum stress in the heart: Insights into mechanisms and drug targets. British Journal of Pharmacology. 2018;**175**:1293-1304. DOI: 10.1111/bph.13888

[66] Jin JK, Blackwood EA, Azizi K, Thuerauf DJ, Fahem AG, Hofmann C, et al. ATF6 decreases myocardial ischemia/reperfusion damage and links ER stress and oxidative stress signaling pathways in the heart. Circulation Research. 2017;**120**:862-875. DOI: 10.1161/CIRCRESAHA.116.310266

[67] Ye S, Zhou X, Chen P, Lin JF. Folic acid attenuates remodeling and dysfunction in the aging heart through the ER stress pathway. Life Sciences. 2021;**264**:118718. DOI: 10.1016/j.lfs.2020.118718

[68] Xu J, Zhou Q, Xu W, Cai L. Endoplasmic reticulum stress and diabetic cardiomyopathy. Experimental Diabetes Research. 2012;**2012**:827971. DOI: 10.1155/2012/827971

[69] Yang L, Zhao D, Ren J, Yang J. Endoplasmic reticulum stress and protein quality control in diabetic cardiomyopathy. Biochimica et Biophysica Acta. 2015;**1852**:209-218. DOI: 10.1016/j.bbadis.2014.05.006

[70] Nakamura M, Sadoshima J. Mechanisms of physiological and pathological cardiac hypertrophy. Nature

Reviews. Cardiology. 2018;**15**(7):387-407. DOI: 10.1038/s41569-018-0007-y

[71] Zhang Y, Chen W, Wang Y. STING is an essential regulator of heart inflammation and fibrosis in mice with pathological cardiac hypertrophy via endoplasmic reticulum (ER) stress. Biomedicine & Pharmacotherapy. 2020;**125**:110022. DOI: 10.1016/j.biopha.2020.110022

[72] Liu S, Sun WC, Zhang YL, Lin QY, Liao JW, Song GR, et al. SOCS3 negatively regulates cardiac hypertrophy via targeting GRP78-mediated ER stress during pressure overload. Frontiers in Cell and Development Biology. 2021;**9**:629932. DOI: 10.3389/fcell.2021.629932

[73] Richards DA, Aronovitz MJ, Calamaras TD, Tam K, Martin GL, Liu P, et al. Distinct phenotypes induced by three degrees of transverse aortic constriction in mice. Scientific Reports. 2019;**9**(1):5844. DOI: 10.1038/s41598-019-42209-7

[74] Okada K, Minamino T, Tsukamoto Y, Liao Y, Tsukamoto O, Takashima S, et al. Prolonged endoplasmic reticulum stress in hypertrophic and failing heart after aortic constriction: Possible contribution of endoplasmic reticulum stress to cardiac myocyte apoptosis. Circulation. 2004;**110**(6):705-712. DOI: 10.1161/01.CIR.0000137836.95625.D4

[75] Chadwick SR, Lajoie P. Endoplasmic reticulum stress coping mechanisms and lifespan regulation in health and diseases. Frontiers in Cell and Development Biology. 2019;**7**:84. DOI: 10.3389/fcell.2019.00084

[76] Hipp MS, Kasturi P, Hartl FU. The proteostasis network and its decline in ageing. Nature Reviews. Molecular Cell Biology. 2019;**20**(7):421-435. DOI: 10.1038/s41580-019-0101-y

[77] Liu M, Dudley SC Jr. The role of the unfolded protein response in arrhythmias. Channels (Austin, Tex.). 2018;**12**(1):335-345. DOI: 10.1080/19336950.2018.1516985

[78] Sirish P, Diloretto DA, Thai PN, Chiamvimonvat N. The critical roles of proteostasis and endoplasmic reticulum stress in atrial fibrillation. Frontiers in Physiology. 2022;**12**:793171. DOI: 10.3389/fphys.2021.793171

[79] Nakamura S, Numata G, Yamaguchi T, Tokiwa H, Higashikuni Y, Nomura S, et al. Endoplasmic reticulum stress-activated nuclear factor-kappa B signaling pathway induces the upregulation of cardiomyocyte dopamine D1 receptor in heart failure. Biochemical and Biophysical Research Communications. 2022;**637**:247-253. DOI: 10.1016/j.bbrc.2022.11.031

[80] Hamilton S, Terentyev D. ER stress and calcium-dependent arrhythmias. Frontiers in Physiology. 2022;**13**:1041940. DOI: 10.3389/fphys.2022.1041940

[81] Pitsch M, Kant S, Mytzka C, Leube RE, Krusche CA. Autophagy and endoplasmic reticulum stress during onset and progression of arrhythmogenic cardiomyopathy. Cell. 2021;**11**(1):96. DOI: 10.3390/cells11010096

[82] Ni L, Zhou C, Duan Q, Lv J, Fu X, Xia Y, et al. β-AR blockers suppresses ER stress in cardiac hypertrophy and heart failure. PLoS One. 2011;**6**(11):e27294. DOI: 10.1371/journal.pone.0027294

[83] Yan J, Wang J, He JC, Zhong Y. Sirtuin 1 in chronic kidney disease and therapeutic potential of targeting sirtuin 1. Frontiers in Endocrinology (Lausanne). 2022;**13**:917773. DOI: 10.3389/fendo.2022.917773

[84] Filadi R, Theurey P, Pizzo P. The endoplasmic reticulum-mitochondria coupling in health and disease: Molecules, functions and significance. Cell Calcium. 2017;**62**:1-15. DOI: 10.1016/j.ceca.2017.01.003

[85] Gardner BM, Pincus D, Gotthardt K, Gallagher CM, Walter P. Endoplasmic reticulum stress sensing in the unfolded protein response. Cold Spring Harbor Perspectives in Biology. 2013;**5**:a013169. DOI: 10.1101/cshperspect.a013169

[86] Giacomello M, Pellegrini L. The coming of age of the mitochondria-ER contact: A matter of thickness. Cell Death and Differentiation. 2016;**23**:1417-1427. DOI: 10.1038/cdd.2016.52

[87] Hu D, Barajas-Martínez H, Burashnikov A, Panama BK, Cordeiro JM, Antzelevitch C. Mechanisms underlying atrial-selective block of sodium channels by Wenxin Keli: Experimental and theoretical analysis. International Journal of Cardiology. 2016;**207**:326-334. DOI: 10.1016/j.ijcard.2016.01.016

[88] Wu A, Zhai J, Zhang D, Lou L, Zhu H, Gao Y, et al. Effect of wenxin granule on ventricular remodeling and myocardial apoptosis in rats with myocardial infarction. Evidence-based Complementary and Alternative Medicine. 2013;**2013**:967986. DOI: 10.1155/2013/967986

[89] Tu T, Li B, Li X, Zhang B, Xiao Y, Li J, et al. Dietary ω-3 fatty acids reduced atrial fibrillation vulnerability via attenuating myocardial endoplasmic reticulum stress and inflammation in a canine model of atrial fibrillation. Journal of Cardiology. 2022;**79**:194-201. DOI: 10.1016/j.jjcc.2021.08.012

Chapter 8

# Targeting Endoplasmic Reticulum and Mitochondrial Dynamics to Combat Triple-Negative Breast Cancer

*Priyanka Menon Kunnel and Bibu John Kariyil*

## Abstract

Triple negative breast cancer (TNBC) is a cancer that is aggressive with short survival rate. In comparison to other breast cancer subtypes, TNBC tumors are bigger, more chemo resistant, highly proliferative, and usually more abundant in stem and immune cells. These modifications are functionally dependent on a high-quality endoplasmic reticulum and mitochondrial pool. Endoplasmic reticulum and mitochondrial health are monitored and enhanced on a regular basis via endoplasmic reticulum and mitochondrial dynamics. The role of endoplasmic reticulum and mitochondrial dynamics in tumor growth and metastasis has been highlighted by recent advances in understanding the endoplasmic reticulum and mitochondrial dynamics in TNBC. This chapter examines the current knowledge of endoplasmic reticulum and mitochondrial dynamics in TNBC.

**Keywords:** endoplasmic reticulum dynamics, endoplasmic reticulum transmembrane proteins, mitochondrial dynamics, molecular mechanisms, apoptosis, triple-negative breast cancer

## 1. Introduction

Triple-negative breast cancer (TNBC) is a subgroup of breast tumors that does not have estrogen receptor (ER), progesterone receptor (PR), and human epidermal growth factor receptor 2 ($HER_2$) that are commonly found in breast cancer. TNBC cells undergo significant molecular alterations at the cellular level. They are highly proliferative, more chemo resistant, larger in size, and more aggressive than other types of breast cancer (BC) cells. Compared to other BC subtypes, this subtype has a higher concentration of stem cells and lymphocytes. Porporato et al. reviewed that all stages of cancer development, including proliferation, malignant transformation, and progression, require mitochondria [1]. They not only supply sufficient energy but also protect TNBC cells from excessive amounts of reactive oxygen species (ROS). Reactive oxygen species improve TNBC cells metastatic potential in the early stages of development. High amounts of ROS, on the other hand, cause ROS-induced apoptosis, which is harmful to cancer growth [2]. To survive significant ROS-related damage, TNBC cells must have powerful defensive mechanisms. Mitochondrial dynamics are crucial in maintaining this equilibrium [3].

The endoplasmic reticulum (ER) is an intracellular organelle that is dynamic in both function and structure. It plays a variety of roles in cellular homeostasis, development, and stress tolerance. It is the organelle in charge of protein folding, translocation, and posttranslational modification. Oxidative stress, altered glycosylation, nutrient deprivation, calcium depletion, DNA damage, and energy disturbance are all caused by physiological, biochemical, and pathological stimuli in the ER, resulting in ER stress and the build-up of unfolded or misfolded proteins in the ER. In order to survive, the cells must overcome ER dysfunction and ER stress. Apoptosis can result from unresolved ERS [4].

In TNBC cells, pharmacological aggravation of ER stress causes significant cell death, and this method is even effective in multidrug-resistant forms [5]. The ER stress process has been connected to autophagy, a cellular function that appears to be important for general cell homeostasis, cancer, and chemoresistance. Autophagy and ER stress are linked: several agents that aggravate ER stress cause an increase in autophagic activity; on the other hand, there is evidence that blocking autophagy promotes ER stress [6]. *In vitro* and *in vivo* augmenting of ER stress and simultaneously inhibiting autophagy may result in effective TNBC death [7].

## 2. Endoplasmic reticulum dynamics and role of ER in TNBC

### 2.1 Endoplasmic reticulum dynamics

ER stress promotes an increase in transcription of p53 unregulated modulator of apoptosis (PUMA), Bcl2-like11 (BIM), BH3-only proteins, and NADPH oxidase activator (NOXA) due to an imbalance between anti- and pro-apoptotic Bcl-2 proteins. ER stress promotes interactions between Bax and PUMA. This results in the release of cytochrome c and caspase-dependent cleavage of p53 resulting in apoptosis [8]. ER stress in tumor cells may restore homeostasis. It also makes the surrounding environment more conducive to tumor survival and proliferation [9]. Malnutrition, hypoxia, pH fluctuations, and poor vascularization are stressful situations that might inhibit tumor cell development and activate the unfolded protein response (UPR). ER stress is caused by both nutritional deficiency [10] in tumor cells and nutrient excess in normal conditions [11]. High proliferation rates of cancer cells necessitate higher ER protein folding, assembly, and transport activities. This might result in physiological ER stress. This response is cytoprotective and plays a role in tumor development as well as adaptation to harsh conditions [12].

Inositol-requiring enzyme 1α (IRE1α), pancreatic ER kinase-like ER kinase (PERK), and activating transcription factor 6 (ATF6) localized in the ER, are the three ER stress signaling branches involved in tumorigenesis. X-box binding protein (XBP1) aids in cancer progression and is increased in breast cancer [13]. The ER resident chaperone calreticulin is found on the cell surface of tumor cells and has been linked to immunogenic cell death and calreticulin localization on tumor cell surfaces. This link could be due to ER stress in tumor cells [14]. ER chaperone and UPR components are over-expressed in breast cancers.

The ER stress response, on the other hand, is also directly implicated in pro-apoptotic pathways in both UPR-dependent and UPR-independent ways [15]. To activate IRE1α, the cytosolic domain of IRE1α interacts with the Bax/Bak apoptotic pathway [16]. EI24/PIG8, an ER-localized Bcl2-binding protein, inhibits breast cancer invasiveness via modulating Bcl-2 function [17]. Bim is also involved in the death of MCF-7 cells generated from breast cancer by activating ER stress-induced apoptosis [18]. Another potential anti-tumor technique is to activate the CHOP-GADD34 axis [19]. In ER stress-exposed cells, CHOP causes cell death by increasing protein synthesis and oxidation [20].

### 2.2 Unfolded protein response

Depending on the cell state, the UPR can be both cytoprotective and cytotoxic. The UPR's objective is to keep the ER folding environment in check when it is under stress. Tumor cells will die if UPR fails to restore ER equilibrium under prolonged ER stress. In combination with induced tumor dormancy, the UPR can shield tumor cells against apoptosis, allowing for tumor regeneration once favorable conditions have been restored [21].

By dissociating Grp78/binding immunoglobulin protein (Bip), a key chaperone protein, from three membrane-bound ER stress sensors, comprising ATF6, PERK, and IRE1α, cells aim to maintain normal folding processes in the ER [22]. Following the separation of detecting proteins from Grp78/Bip, these sensors are activated in order, with PERK being the first, blocking general protein synthesis by phosphorylating eIF2α [23]. During cellular stress, these activities also cause the transcription factor NF-κB to be inhibited. Another transcription factor triggered by translocation to the Golgi apparatus is ATF6, which is cleaved and the transcription factor in its active form is released to regulate gene expression [24]. Following IRE-1 activation and downregulation, splicing of XBP1 occurs, and the spliced XBP1 protein translocates to the nucleus, where it activates the transcription of chaperones involved in protein folding and secretion or ER-associated protein degradation (ERAD) [25]. Rapid tumor growth and insufficient vascularization occur during carcinogenesis, resulting in microenvironmental stress [26].

### 2.3 Role of ER in TNBC

Since cancer cells are constantly dividing, they are affected by lack of nutrition and oxygen, as well as by lack of vascularization. As a result, cancer alters the expression patterns of ER resident proteins. On tumors, ER stress has two effects. It has adaptive meaning in the sense that it promotes tumor growth. Second, it possesses cytotoxic properties that cause apoptosis. Cancer cells activate UPR to adapt to their surroundings, and macrophages secrete cytokines, growth factors, and angiogenic substances to generate more favorable microenvironments for cancer cell proliferation and invasiveness [27]. Cancer cells use NF-κB pathways to promote cyclooxygenase-2 expression during ER stress, which has antiapoptotic effects. It also maintains IL-8 production in human epithelial cells and enhances pro-inflammatory NF-κB activation via CHOP [28].

Apoptosis is caused by a variety of pathways. One such pathway is ER stress. The caspase-12 family of proapoptotic cysteine proteases is linked with the ER membrane and plays a key role in ER stress-induced apoptosis, although it is not activated by non-ER stimuli [29]. Vascular endothelial growth factor (VEGF) promotes endothelial cell proliferation and angiogenesis by increasing Grp78 expression on the endothelial surface. Through mitogen-activated protein kinase (MAPK) signaling, Grp78 knockdown reduces endothelial cell growth [30]. The action of P38MAPK keeps cells in a $G_0$-like quiescent state [31]. In chicken embryo chorioallantois membrane system and subcutaneous xenograft models, PERK-elf2α also pauses cell development at $G_0$/$G_1$ and prevents carcinogenesis [32].

#### *2.3.1 Glucose-regulated protein 78/binding immunoglobulin protein*

Grp78, an ER chaperone protein, is one of the cancer cells' most active components and is overexpressed in a variety of malignancies [33]. It is been considered as a chaperone protein that helps cancer cells adapt to hypoxic conditions and as a resistance protein to anti-cancer drugs [34]. In cancer systems, Grp78 affects cell apoptosis, proliferation, invasion, inflammation, and immunity [35]. It has recently been discovered to play a role in carcinogenesis, metastasis, and angiogenesis [36]. Grp78, through physical and functional interactions with BIK in the ER, suppresses BIK-mediated apoptosis and confers resistance to estrogen starvation-induced apoptosis in human breast cancer cells [37]. Grp78 is primarily found inside the ER, although it may be translocated to the surface of tumor cells during ER stress [38]. In addition to the ER, some Grp78 is found in the cytosol, nucleus, and mitochondria during ER stress [39]. A noteworthy prospective anticancer treatment is to block Grp78 translocation. Grp78, present on the surface of cell was found in receptor-positive BT474 breast cancer cells but not in triple-negative MDA-MB-468 cells. The absence of ER, PR, and HER2 receptors in this TNBC was linked to Grp78 negative expression on tumor cells of breast cancer [40]. GRP78 expression was also found to be low in TNBC cell lines, such as MDA-MB-231, by others [41]. Grp78 expression was found to be substantially linked to TNBC invasiveness, distant metastasis, and proliferation. Also, patients with Grp78 expression were found to have shorter overall survival (OS) and disease-free survival (DFS). Furthermore, elevated Grp78 expression was linked to disease-free survival (DFS) in TNBC patients [42].

## 3. Role of ER transmembrane proteins in TNBC

### 3.1 Pancreatic endoplasmic reticulum kinase-like endoplasmic reticulum kinase

PERK/eIF2 regulates tumor initiation and survival, making it easier for cells to adapt to varied circumstances including hypoxia and oxidative stress [43]. Tumor cells multiply quickly, resulting in the creation of new blood vessels and, eventually, a link to the microenvironment and nutrition restriction. Cytotoxic circumstances result from increased demand for glucose and oxygen. Reactive oxygen species (ROS) are produced when the generation of ATP and NADPH in a reducing equivalent form is disrupted. ROS build up in the mitochondria, causing ER stress to activate [44]. A cellular stress sensor in the ER responds to changes in nutritional shortage, which has been linked to cancer. PERK is a trans-ER membrane serine/threonine protein kinase with an ER luminal domain at the N-terminus and a cytoplasmic protein kinase domain at the C-terminus [45]. The transcription factors Nrf2 transcription factor [46] and elF2 are phosphorylated by PERK. The phosphorylation of elF2 suppresses most transcript translation while promoting the translation of a few mRNAs, such as the transcription factor ATF4 [47]. PERK phosphorylates Nrf2, which is then liberated from an inhibiting E3 ligase complex and translocated into the nucleus, where it generates enzymes that reduce intracellular ROS [48]. When hypoxia occurs in tumors, the transcription regulator HIF1 is stabilized, and the whole branch of the UPR, namely PERK, is fully activated, resulting in the phosphorylation of elf2, ATF4, and GADD34. Although elf2 phosphorylation reduces overall protein synthesis, ATF4, a transcription factor, is linked to cancer cell growth and survival despite food restriction via amino acid synthesis [49]. TNBC viability and proliferation were reduced *in vivo* by CCT020312, a specific eIF2/PERK activator that activated the PERK/eIF2/ATF4/CHOP pathway while inactivating the AKT/mTOR pathway [50]. In luminal androgen receptor (LAR), TNBC ER stress reduced androgen receptor (AR) expression at the transcriptional level via PERK/eIF2/ATF4 signaling. ATF4 also inhibits AR promoter activity by binding to the promoter region of AR from −2813 to −2486 nt and from −2084 to −1742 nt [51]. In MDA-MB-468 and T47D cell lines, PERK-dependent signaling is involved in tumor initiation and expansion to maintain redox homeostasis and drive tumor growth [52].

### 3.2 Inositol-requiring enzyme 1α/X-box binding protein

In cells and tissues, an ER transmembrane sensor called IRE1α protects against ER stress. IRE1 is activated during ER stress by autophosphorylation and oligomerization, which causes its endoribonuclease to cleave and start splicing the XBP1 mRNA [53]. IRE1α-dependent mRNA decay (RIDD), which is different from XBP1 splicing, aids in the restoration of ER homeostasis. This is done by targeting mRNAs encoding secretory proteins. IRE1α RNase activity controls the activity of RIDD [54]. In the UPR, the IRE1-XBP1 pathway has also been suggested to play a pro-survival role [53]. The UPR, on the other hand, causes cellular death in the presence of persistent and uncompensated stress. [55]. IRE1-TRAF2-ASK is another pathway that has been proposed. Phosphorylation of IRE1 causes it to bind to tumor-necrosis factor receptor-associated factor 2 (TRAF2) and activate apoptosis signal-regulating kinase (ASK1), resulting in JNK and p38 activation and ER-stressed caused cell death [56]. By directly activating procaspase-4, the IRE1 and TRAF2 pathways are also engaged in mitochondria-independent apoptosis [57]. TNBCs have an elevated basal level of endoplasmic reticulum stress. There is activation of the XBP1 branch of the UPR, a significant cellular stress response system in the tumor microenvironment [58].

### 3.3 Inositol 1,4,5 triphosphate receptors

Calcium ions play a key in the development of metastases. The cytosolic $Ca^{2+}$ concentration increases by a factor of 5 to 10 when $Ca^{2+}$ channels activate (from 100 nM to 500–1000 nM). Part of this calcium signaling is generated inside the cell by inositol (1,4,5)-triphosphate (IP3), which has three identified subtypes: IP3R1, IP3R2, and IP3R3. Each subtype has its own calcium release signature: IP3R2 produces strong $Ca^{2+}$ oscillations, IP3R1 produces milder oscillations, while IP3R3 produces monophasic transients [59].

The ER is the cell's principal intracellular $Ca^{2+}$ storing organelle. $Ca^{2+}$ is rapidly transported between the two intracellular organelles due to the tight connection between mitochondria and ER membranes [60]. The IP3R, along with the ryanodine receptors, has been identified as the primary ER $Ca^{2+}$ channel [61]. Massive and long-term $Ca^{2+}$ overload in the mitochondria can open the permeability transition pore [62]. As a result, proapoptotic and caspase-activating substances in mitochondria, such as cytochrome c, are released into the cytoplasm. By binding to IP3R, cytochrome c in the cytoplasm exacerbates $Ca^{2+}$ release, avoiding $Ca^{2+}$ −dependent regulation of the receptor and increasing caspase activation to promote apoptosis [63].

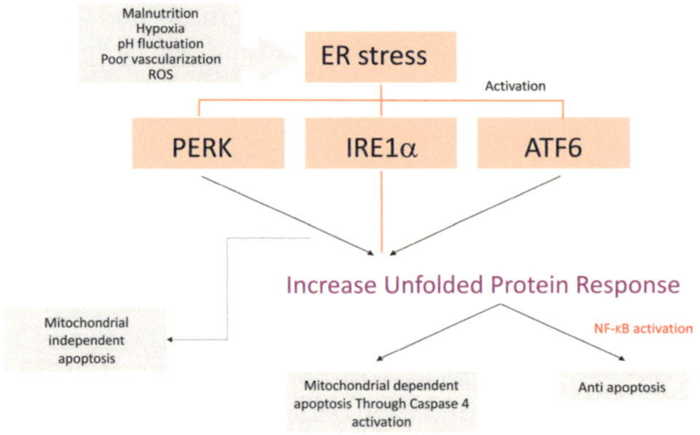


**Figure 1.**
*Endoplasmic reticular stress activates the ER stress signaling branches that cause an increase in unfolded protein response. Depending on the cell status, UPR leads to either apoptosis or anti-apoptosis.*

In breast cancer tissue, IP3R3 is the most highly expressed subtype. Furthermore, invasive breast cancer tissue, such as TNBCs, has much higher levels of IP3R3 and IP3R1 expression than non-tumor tissue. Because its expression is linked to the severity of BC, IP3R3 can be regarded as a sign of aggressiveness in BC [59].

### 3.4 STAT 3

STAT3 is constitutively activated and overexpressed in TNBC cells. By regulating the expression of its downstream target genes, it contributes to cell survival, cell cycle progression, migration, invasion, anti-apoptosis, chemoresistance, immunosuppression, and stem cell self-renewal and differentiation [64]. STAT3, an oncogenic transcription factor, has been found to be abundant in ER and MAM lately. Reduced ER-mitochondria $Ca^{2+}$ transport and IP3R3 degradation mediated by the IP3R3/STAT3 association increase cell resistance to apoptosis in constitutively active STAT3 (**Figure 1**) [65].

## 4. Mitochondrial dynamics in TNBC

Mitochondrial dynamics is a mechanism that regulates the quality of the mitochondrial pool. In simple terms, it is a dynamic flow of mitochondria that includes divisions and assembly, allowing for the destruction or repair of malfunctioning parts. Mitochondrial dynamics thus serve as the first line of defense against ROS-induced mitochondrial damage. Given that TNBC cells have a high quantity of ROS, it is assumed that this process is critical for TNBC survival.

Changing mitochondrial dynamics helps cancer cells survive because they regulate a delicate balance between delivering energy and apoptosis. Many cancer tissues, including lung, glioma, neuroblastoma, colorectal, pancreatic, and melanoma, contain fragmented mitochondria [66]. Similarly, mitochondrial fission is significantly enhanced in TNBC clinical samples and is associated with poorer TNBC patient survival. Drp1 was upregulated and Mfn1 was downregulated in TNBC tumor tissues, according to immunohistochemistry (IHC) labeling. The results are further supported by cancer cell labeling and the relative mRNA expression ratio of tumor/

peritumor determined by qRTPCR [67]. Hypoxia also increases Drp1 expression in TNBC MDA-MB-231 cells, but not in ER-positive MCF7 cells [68]. TNBC tumors are more hypoxic than other BC subtypes, therefore this pattern is particularly intriguing [69]. A mutation in the tumor stressor gene TP53 is also found in the majority of TNBC tumors [70].

Increase in fission offers TNBC tumors with an optimal environment for development, given that knockout mice without p53 (Trp53) are able to survive longer in hypoxia environments because tumorgenesis levels are lower than in normoxic situations. Fission has been identified as a critical stage in the release of cytochrome c and, as a result, apoptosis progression. Fission-dependent apoptosis, on the other hand, appears to be context and cell-type dependent. Mitochondrial fission boosted TNBC cell growth *in vitro* and *in vivo* by suppressing apoptosis and enhancing proliferation through a notch-dependent mechanism [71].

### 4.1 Mitochondrial fusion and fission

Mitochondria are double membrane organelles consisting of an outer mitochondrial membrane (OMM), an inner mitochondrial membrane (IMM), and an intermembrane space between the two membranes. They are not static organelles, but rather a dynamic pool that is constantly undergoing fusion and fission. This dynamic process is essential for mitochondrial health and cellular adaptation to environmental changes.

Mitochondrial fusion occurs when two mitochondria fuse to form one mitochondrion. Mitofusin 1 (Mfn1) and mitofusin 2 (Mfn2), as well as optic atrophy protein 1 (OPA1) are critical proteins in mitochondrial fusion. Mfn1 and 2 are homodimers and heterodimers that are found near the OMM. The mitofusins are in control of OMM fusion, while OPA1 is in charge of IMM fusion. Furthermore, OPA1 interacts with Mfn2 during fusion, and interruption of this mechanism results in mitochondrial fission. ROS induce mutations in mtDNA, which results in altered respiratory functions in mitochondria. As a result, mitochondrial fusion is critical because it allows for the interchange of metabolites and gene products between fusing mitochondria, improving overall respiratory performance. Similarly, mitochondrial fusion impediment is linked to a reduction in mitochondrial function [72]. Fusion of mitochondrion and promotion of docking is the primary function of mitofusins. The oligomerization of the GTPase domains of MFNs is essential for the tethering of two OMMS, this oligomerization also requires GTP hydrolysis. When GTP is bound and hydrolyzed, the GTPase domains undergo a conformational shift that causes them to oligomerize, allowing the two mitochondria to dock at the two outer membranes and fuse. Ubiquitination, deacetylation, and phosphorylation regulate MFN 1's activity. The extracellular signal-regulated kinase (ERK) phosphorylates Mfnl in the HRI domain, which inhibits mitochondrial fusion and causes death. In situations of glucose deprivation, however, deacetylation of Mfn1 by histone deacetylase 6 (HDAC6) leads to its activation and facilitation of fusion. JNK phosphorylates MFN2 in response to cellular stress, which recruits E3 ubiquitin ligase, which ubiquitinates MFN2 and causes its proteasomal destruction. MFN2 degradation leads to mitochondrial fragmentation and an increase in apoptotic cell death. S-OPA1 and L-OPA1 combination is essential for IMM fusion. Similarly, discovered that L-OPA1 and S-OPA1 collaborate to promote fusion activity in liposomes, resulting in efficient and rapid membrane pore opening. Other research, on the other hand, has found that only L-OPA1 is required to stimulate IMM fusion [73]. A phospholipid cardiolipin, IMM component is required for IMM fusion.

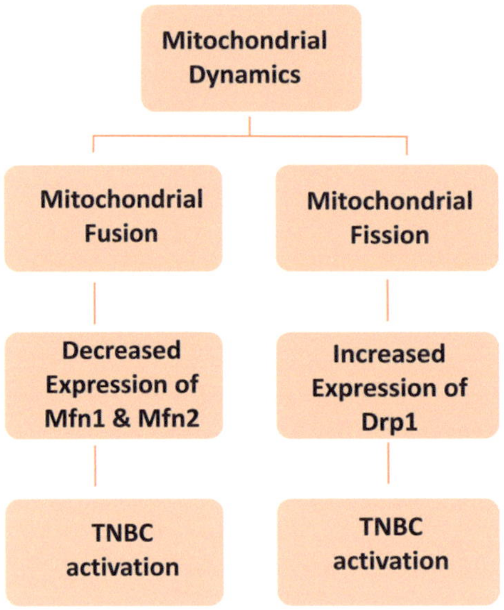


**Figure 2.**
*Mfn1 and Mfn2 are the membrane proteins involved in mitochondrial fusion, its downregulation activates TNBC cells. Drp1 is the main membrane protein involved in mitochondrial fission. Its upregulation activates TNBC cells.*

Mitochondrial fission occurs when one mitochondrion splits into two daughter mitochondria. Dynamin-related proteins (DRPs) and cofactors mediate mitochondrial fission. Constriction of mitochondria occurs where ER binds to mitochondria. At the ER–mitochondrial interaction point, Drp1 is recruited to OMM. Drp1 binding site and constriction are indicated by contact between ER tubules and mitochondria. Drp1 binds to the ER binding site on the OMM and interacts with adaptor proteins such as mitochondrial fission factor (MFF), mitochondrial dynamics protein, and cardiolipin (CL) once activated. Drpl undergoes oligomerization when the fission machinery is assembled, forming a ring around the OMM. Drp1, MFF, Mid49/51, and Mdv1-dependent GTP hydrolysis provides energy for scission completion and constriction [73].

Various cellular stresses, ROS, hypoxia, posttranslational modification, etc., regulate mitochondrial dynamics.

When these spatiotemporal events are disrupted, the consequence is either a fragmented network with many small round mitochondria or a hyperfused network with elongated mitochondria (**Figure 2**).

## 5. Mitochondria-associated ER membranes

Mitochondria-associated ER membranes (MAMs) function as signaling hubs that regulate ER and mitochondrial activity. They regulate lipid metabolism, $Ca^{2+}$ homeostasis, mitochondrial function and cell death. Bcl2, p53 tumor suppressor, etc., are the cell-death pathway protein involved in MAM [74]. Also, proteins like Mfn2 and Drp1 are enriched in MAM. Mfn2 is present on both OMM and ER. Mfn2 along with Mfn1 on OMM establish ER–mitochondrial interactions [75]. Rapid $Ca^{2+}$ exchange between mitochondria and ER also determines mitochondrial bioenergetics and cell fate. An oxidizing enzyme endoplasmic reticulum oxidoreductin 1-α (ERO1-α) is enriched in

MAM. ERO1-α affects ER redox homeostasis, ER $Ca^{2+}$ flow, and consequent mitochondrial $Ca^{2+}$ build-up. It also modulates oxidative folding. ERO1-α expression is correlated with programmed cell death ligand 1 (PD-L1) in TNBC [76]. It is also an indicator of poor prognosis in breast cancer.

When MAMs are disrupted, a variety of cellular processes, such as apoptosis, inflammation, and autophagy, become dysfunctional. These cellular processes are very important in the pathophysiology of TNBC [74].

## 6. Conclusion

The significance of the ER stress response and dynamic transformations of mitochondria, which regulate cell fate decisions in carcinogenesis and cancer resistance, is being clarified by increasing evidence. Understanding how these events are controlled from not only a molecular perspective but also a biological perspective is essential to comprehending a variety of human disorders. The search for novel medications to control these occurrences is an ongoing process.

## Conflict of interest

The authors declare no conflict of interest.

## Author details

Priyanka Menon Kunnel and Bibu John Kariyil*
Faculty of Veterinary and Animal Sciences, College of Veterinary and Animal Sciences, Kerala Veterinary and Animal Sciences University, Thrissur, Kerala, India

*Address all correspondence to: bibujohn@kvasu.ac.in

## References

[1] Porporato PE, Filigheddu N, Pedro JMBS, Kroemer G, Galluzzi L. Mitochondrial metabolism and cancer. Cell Research. 2018;**28**:265-280. DOI: DOI.10.1038/cr.2017.155

[2] Cheng CT, Kuo CY, Ouyang C, Li CF, Chung Y, Chan DC, et al. Metabolic stress-induced phosphorylation of KAP1 Ser473 blocks mitochondrial fusion in breast cancer cells. Cancer Research. 2016;**76**:5006-5018. DOI: DOI. 10.1158/0008-5472.CAN-15-2921

[3] Kashatus DF. The regulation of tumor cell physiology by mitochondrial dynamics. Biochemical Biophysical Research Communication. 2018;**2018**(500):9-16. DOI: DOI. 10.1016/j.bbrc.2017.06.192

[4] Tabas I, Ron D. Integrating the mechanisms of apoptosis induced by endoplasmic reticulum stress. Nature Cell Biology. 2011;**13**(3):184-190. DOI: 10.1038/ncb0311-184

[5] de Ruijter TC, Veeck J, de Hoon JP, van Engeland M, Tjan-Heijnen VC. Characteristics of triple-negative breast cancer. Journal of Cancer Research and Clinical Oncology. 2011;**137**:183-192. DOI: 10.1007/s00432-010-0957-x

[6] Schönthal AH. Endoplasmic reticulum stress and autophagy as targets for cancer therapy. Cancer Letters. 2009;**275**:163-169. DOI: 10.1016/j.canlet.2008.07.005

[7] Thomas S, Sharma N, Golden EB, Cho H, Agarwal P, Gaffneyy KJ, et al. Preferntial killing of triple-negative breast cancer cells in vitro and in vivo when pharmacological aggravators of endoplasmic reticulum stress are combined with autophagy inhibitors. Cancer Letters. 2012;**325**:63-71. DOI: 10.1016/j.canlet.2012.05.030

[8] Rao RV, Niazi K, Mollahan P, Mao X, Crippen D, Poksay KS, et al. Coupling endoplasmic reticulum stress to the cell-death program: A novel HSP90-independent role for the small chaperone protein p23. Cell Death and Differentiation. 2006;**13**(3):415-425. DOI: 10.1038/sj.cdd.4401761

[9] Martinon F. Targeting endoplasmic reticulum signaling pathways in cancer. Acta Oncologica. 2012;**51**(7):822-830. DOI: 10.3109/0284186X.2012.689113

[10] Ozcan U, Cao Q, Yilmaz E, Lee AH, Iwakoshi NN, Ozdelen E, et al. Endoplasmic reticulum stress links obesity, insulin action, and type 2 diabetes. Science. 2004;**306**(5695): 457-461. DOI: 10.1126/science.1103160

[11] Tsai YC, Weissman AM. The unfolded protein response, degradation from the endoplasmic reticulum, and cancer. Genes and Cancer. 2010;**1**(7): 764-778. DOI: 10.1177/1947601910383011

[12] Healy SJ, Gorman AM, Mousavi-Shafaei P, Gupta S, Samali A. Targeting the endoplasmic reticulum-stress response as an anticancer strategy. European Journal of Pharmacology. 2009;**625**(1-3):234-246. DOI: 10.1016/j.ejphar.2009.06.064

[13] Koong AC, Chauhan V, Romero-Ramirez L. Targeting XBP-1 as a novel anti-cancer strategy. Cancer Biology and Therapy. 2006;**5**(7):756-759. DOI: DOI.10.4161/cbt.5.7.2973

[14] Garg AD, Krysko DV, Verfaillie T, Kaczmarek A, Ferreira GB, Marysael T, et al. A novel pathway combining

calreticulin exposure and ATP secretion in immunogenic cancer cell death. EMBO Journal. 2012;**31**:1062-1079. DOI: 10.1038/emboj.2011.497

[15] Moenner M, Pluquet O, Bouchecareilh M, Chevet E. Integrated endoplasmic reticulum stress responses in cancer. Cancer Research. 2007;**67**(22):10631-10634. DOI: 10.1158/0008-5472.CAN-07-1705

[16] Hetz C, Bernasconi P, Fisher J, Lee AH, Bassik MC, Antonsson B, et al. Proapoptotic BAX and BAK modulate the unfolded protein response by a direct interaction with IRE1α. Science. 2006;**312**(5773):572-576. DOI: 10.1126/science.1123480

[17] Zhao X, Ayer RE, Davis SL, Ames SJ, Florence B, Torchinsky C, et al. Apoptosis factor EI24/PIG8 is a novel endoplasmic reticulum–localized Bcl-2–binding protein which is associated with suppression of breast cancer invasiveness. Cancer Research. 2005;**65**(6):2125-2129. DOI: 10.1158/0008-5472.CAN-04-3377

[18] Puthalakath H, O'Reilly LA, Gunn P, Lee L, Kelly PN, Huntington ND, et al. ER stress triggers apoptosis by activating BH3-only protein Bim. Cell. 2007;**129**(7):1337-1349. DOI: 10.1016/j.cell.2007.04.027

[19] Dalton LE, Clarke HJ, Knight J, Lawson MH, Wason J, Lomas DA, et al. The endoplasmic reticulum stress marker CHOP predicts survival in malignant mesothelioma. British Journal of Cancer. 2013;**108**(6):1340-1347. DOI: DOI10.1038/bjc.2013.66

[20] Marciniak SJ, Yun CY, Oyadomari S, Novoa I, Zhang Y, Jungreis R, et al. CHOP induces death by promoting protein synthesis and oxidation in the stressed endoplasmic reticulum. Genes and Development. 2004;**18**(24):3066-3077. DOI: 10.1101/gad.1250704

[21] Dong D, Stapleton C, Luo B, Xiong S, Ye W, Zhang Y, et al. A critical role for GRP78/BiPinthetumormicroenvironment for neovascularization during tumor growth and MetastasisRole of GRP78 in the tumor microenvironment. Cancer Research. 2011;**71**(8):2848-2857. DOI: 10.1158/0008-5472.CAN-10-3151

[22] Szegezdi E, Logue SE, Gorman AM, Samali A. Mediators of endoplasmic reticulum stress induced apoptosis. EMBO Reports. 2006;**7**(9):880-885. DOI: 10.1038/sj.embor.7400779

[23] Scheuner D, Song B, McEwen E, Liu C, Laybutt R, Gillespie P, et al. Translational control is required for the unfolded protein response and in vivo glucose homeostasis. Molecular Cell. 2001;**7**(6):1165-1176. DOI: 10.1016/S1097-2765(01)00265-9

[24] Schindler AJ, Schekman R. In vitro reconstitution of ER-stress induced ATF6 transport in COPII vesicles. Proceedings of the National Academy of Sciences. 2009;**106**(42):17775-17780. DOI: 10.1073/pnas.0910342106

[25] Lee AH, Iwakoshi NN, Anderson KC, Glimcher LH. Proteasome inhibitors disrupt the unfolded protein response in myeloma cells. Proceedings of the National Academy of Sciences. 2003;**100**(17):9946-9951. DOI: 10.1073/pnas.1334037100

[26] Vandewynckel YP, Laukens D, Geerts A, Bogaerts E, Paridaens A, Verhelst X, et al. The paradox of the unfolded protein response in cancer. Anticancer Research. 2013;**33**(11):4683-4694

[27] Allavena P, Sica A, Garlanda C, Mantovani A. The Yin-Yang of

tumor-associated macrophages in neoplastic progression and immune surveillance. Immunological Reviews. 2008;**222**(1):155-161. DOI: 10.1111/j.1600-065X.2008.00607.x

[28] Hung JH, Su IJ, Lei HY, Wang HC, Lin WC, Chang WT, et al. Endoplasmic reticulum stress stimulates the expression of cyclooxygenase-2 through activation of NF-κB and pp38 mitogen-activated protein kinase. Journal of Biological Chemistry. 2004;**279**(45):46384-46392. DOI: 10.1074/jbc.M403568200

[29] Nakagawa T, Zhu H, Morishima N, Li E, Xu J, Yankner BA, et al. Caspase-12 mediates endoplasmic-reticulum-specific apoptosis and cytotoxicity by amyloid-β. Nature. 2000;**403**(6765):98-103. DOI: 10.1038/47513

[30] Katanasaka Y, Ishii T, Asai T, Naitou H, Maeda N, Koizumi F, et al. Cancer antineovascular therapy with liposome drug delivery systems targeted to BiP/GRP78. International Journal of Cancer. 2010;**127**(11):2685-2698. DOI: 10.1002/ijc.25276

[31] O'Reilly MS, Holmgren L, Chen C, Folkman J. Angiostatin induces and sustains dormancy of human primary tumors in mice. Nature Medicine. 1996;**2**(6):689-692. DOI: 10.1038/nm0696-689

[32] Ranganathan AC, Zhang L, Adam AP, Aguirre-Ghiso JA. Erratum: Functional coupling of p38-induced up-regulation of BiP and activation of RNA-dependent protein kinasel-like endoplasmic reticulum kinase to drug resistance of dormant carcinoma cells (cancer research (February 1, 2006) 66 (1702-1711)). Cancer Research. 2006;**66**(6):3345. DOI: 10.1158/0008-5472.CAN-05-3092

[33] Li Z, Li Z. Glucose regulated protein 78: A critical link between tumor microenvironment and cancer hallmarks. Biochimica et Biophysica Acta (BBA)-Reviews on CanceActa. 2012;**1826**(1): 13-22. DOI: 10.1016/j.bbcan.2012.02.001

[34] Song MS, Park YK, Lee JH, Park K. Induction of glucose-regulated protein 78 by chronic hypoxia in human gastric tumor cells through a protein kinase C-epsilon/ERK/AP-1 signaling cascade. Cancer Research. 2001;**61**:8322-8330

[35] Kern J, Untergasser G, Zenzmaier C, Sarg B, Gastl G, Gunsilius E, et al. GRP-78 secreted by tumor cells blocks the antiangiogenic activity of bortezomib. Blood. 2009;**114**:3960-3967. DOI: 10.1182/blood-2009-03-209668

[36] Miao YR, Eckhardt BL, Cao Y, Pasqualini R, Argani P, Arap W, et al. Inhibition of established micrometastases by targeted drug delivery via cell surface-associated GRP78. Clinical Cancer Research. 2013;**19**:2107-2116. DOI: 10.1158/1078-0432.CCR-12-2991

[37] Fu Y, Li J, Lee AS. GRP78/BiP inhibits endoplasmic reticulum BIK and protects human breast cancer cells against estrogen starvation-induced apoptosis. Cancer Research. 2007;**67**:3734-3740. DOI: 10.1158/0008-5472.CAN-06-4594

[38] Li Z, Li Z. Glucose regulated protein 78: A critical link between tumor microenvironment and cancer hallmarks. Biochimica et Biophysica Acta. 2012;**1826**:13-22. DOI: 10.1016/j.bbcan.2012.02.001

[39] Ni M, Zhang Y, Lee AS. Beyond the endoplasmic reticulum: A typical GRP78 in cell viability, signalling and therapeutic targeting. Biochemical Journal. 2011;**434**:181-188. DOI: 10.1042/BJ20101569

[40] Raiter A, Yerushalmi R, Hardy B. Pharmacological induction of cell surface

GRP78 contributes to apoptosis in triple negative breast cancer cells. Oncotarget. 2014;**5**(22):11452-11463. DOI: 10.18632/oncotarget.2576. DOI 10.18632/oncotarget.2576

[41] Thomas S, Sharma N, Gonzalez R, Pao P-W, Hofman FM, Chen TC, et al. Repositioning of Verrucosidin, a purported inhibitor of chaperone protein GRP78, as an inhibitor of mitochondrial electron transport chain complex I. PLoS One. 2013;**8**:e65695. DOI: 10.1371/journal

[42] Yang C, Zhang Z, Zou Y, Gao G, Liu L, Xu H, et al. Expression of glucose-regulated protein 78 as prognostic biomarkers for triple-negative breast cancer. Histology and Histopathology. 2019;**35**(6):559-568. DOI: 10.14670/hh-18-185

[43] Bi M, Naczki C, Koritzinsky M, Fels D, Blais J, Hu N, et al. ER stress-regulated translation increases tolerance to extreme hypoxia and promotes tumor growth. EMBO Journal. 2005;**24**:3470-3481. DOI: 10.1038/sj.emboj.7600777

[44] Guzy RD, Hoyos B, Robin E, Chen H, Liu L, Mansfield KD, et al. Mitochondrial complex III is required for hypoxia-induced ROS production and cellular oxygen sensing. Cell Metabolism. 2005;**1**:401-408. DOI: 10.1016/j.cmet.2005.05.001

[45] Harding HP, Zhang Y, Ron D. Protein translation and folding are coupled by an endoplasmic-reticulum-resident kinase. Nature. 1999;**397**:271-274. DOI: 10.1038/16729

[46] Cullinan SB, Zhang D, Hannink M, Arvisais E, Kaufman RJ, Diehl JA. Nrf2 is a direct PERK substrate and effector of PERK-dependent cell survival. Molecular and Cellular Biology. 2003;**23**:7198-7209. DOI: 10.1128/MCB.23.20.7198-7209.2003

[47] Harding HP, Zhang Y, Bertolotti A, Zeng H, Ron D. Perk is essential for translational regulation and cell survival during the unfolded protein response. Molecular Cell. 2000;**5**:897-904. DOI: 10.1016/S1097-2765(00)80330-5

[48] Alam J, Stewart D, Touchard C, Boinapally S, Choi AM, Cook JL. Nrf2, a Cap'n'Collar transcription factor, regulates induction of the heme oxygenase-1 gene. Journal of Biological Chemistry. 1999;**274**:26071-26078. DOI: 10.1074/jbc.274.37.26071

[49] Bi M, Naczki C, Koritzinsky M, Fels D, Blais J, Hu N, et al. ER stress-regulated translation increases tolerance to extreme hypoxia and promotes tumor growth. EMBO Journal. 2005;**24**:3470-3481. DOI: 10.1038/sj.emboj.7600777

[50] Li X, Yu X, Zhou D, Chen B, Li W, Zheng X, et al. CCT020312 inhibits triple-negative breast cancer through PERK pathway-mediated G1 phase cell cycle arrest and apoptosis. Frontiers in Pharmacology. 2020;**11**:737. DOI: 10.3389/fphar.2020.00737

[51] Li X, Zhou D, Cai Y, Yu X, Zheng X, Chen B, et al. Endoplasmic reticulum stress inhibits AR expression via the PERK/eIF2α/ATF4 pathway in luminal androgen receptor triple-negative breast cancer and prostate cancer. NPJ Breast Cancer. 2022;**8**(1):1-3. DOI: 10.1038/s41523-021-00370-1

[52] Bobrovnikova-Marjon E, Grigoriadou C, Pytel D, Zhang F, Ye J, Koumenis C, et al. PERK promotes cancer cell proliferation and tumor growth by limiting oxidative DNA damage. Oncogene. 2010;**29**:3881-3895. DOI: 10.1038/onc.2010.153

[53] Mimura N, Fulciniti M, Gorgun G, Tai YT, Cirstea D, Santo L, et al. Blockade of XBP1 splicing by inhibition of

IRE1alpha is a promising therapeutic option in multiple myeloma. Blood. 2012;**119**:5772-5781. DOI: 10.1182/blood-2011-07-366633

[54] Hollien J, Weissman JS. Decay of endoplasmic reticulum- localized mRNAs during the unfolded protein response. Science. 2006;**313**:104-107. DOI: 10.1126/science.1129631

[55] Urano F, Wang X, Bertolotti A, Zhang Y, Chung P, Harding HP, et al. Coupling of stress in the ER to activation of JNK protein kinases by transmembrane protein kinase IRE1. Science. 2000;**287**:664-666. DOI: 10.1126/science.287.5453.664

[56] Yang W, Tiffany-Castiglioni E, Koh HC, Son IH. Paraquat activates the IRE1/ASK1/JNK cascade associated with apoptosis in human neuroblastoma SH-SY5Y cells. Toxicology Letters. 2009;**191**:203-210. DOI: 10.1016/j.toxlet.2009.08.024

[57] Wu J, Kaufman RJ. From acute ER stress to physiological roles of the unfolded protein response. Cell Death Differerentiation. 2006;**13**:374-384. DOI: 10.1038/sj.cdd.4401840

[58] Chen X, Iliopoulos D, Zhang Q, et al. XBP1 promotes triple-negative breast cancer by controlling the HIF1α pathway. Nature. 2014;**508**:103-107. DOI: 10.1038/nature13119

[59] Foulon A, Rybarczyk P, Jonckheere N, Brabencova E, Sevestre H, Ouadid-Ahidouch H, et al. Inositol (1, 4, 5)-trisphosphate receptors in invasive breast cancer: A new prognostic tool? International Journal of Molecular Sciences. 2022;**23**(6):2962. DOI: 10.3390/ijms23062962

[60] Rizzuto R, Pinton P, Carrington W, Fay FS, Fogarty KE, Lifshitz LM, et al. Close contacts with the endoplasmic reticulum as determinants of mitochondrial $Ca^{2+}$ responses. Science. 1998;**280**:1763-1766. DOI: 10.1126/science.280.5370.1763

[61] Parys JB, De Smedt H. Inositol 1,4,5-trisphosphate and its receptors. Advances in Experimental Medicine and Biology. 2012;**740**:255-279. DOI: 10.1007/978-94-007-2888-2_11

[62] Szabadkai G, Duchen MR. Mitochondria: The hub of cellular $Ca^{2+}$ signaling. Physiology. 2008;**23**:84-94. DOI: 10.1152/physiol.00046.2007

[63] Boehning D, Patterson RL, Sedaghat L, Glebova NO, Kurosaki T, Snyder SH. Cytochrome c binds to inositol (1,4,5) receptors, amplifying calcium dependent apoptosis. Nature Cell Biology. 2003;**12**:1051-1061. DOI: 10.1038/ncb1063

[64] Qin JJ, Yan L, Zhang J, et al. STAT3 as a potential therapeutic target in triple negative breast cancer: A systematic review. Journal of Experimental and Clinical Cancer Research. 2019;**38**:1-6. DOI: 10.1186/s13046-019-1206-z

[65] Avalle L, Camporeale A, Morciano G, Caroccia N, Ghetti E, Orecchia V, et al. STAT3 localizes to the ER, acting as a gatekeeper for ER-mitochondrion $Ca^{2+}$ fluxes and apoptotic responses. Cell Death and Differentiation. 2019;**26**:932-942. DOI: 10.1038/s41418-018-0171-y

[66] Senft D, Ronai ZA. Regulators of mitochondrial dynamics in cancer, in. Current Opinion in Cell Biology. 2016;**39**:43-52. DOI: 10.1016/j.ceb.2016.02.001

[67] Chen L, Zhang J, Lyu Z, Chen Y, Ji X, Cao H, et al. Positive feedback loop between mitochondrial fission and notch signaling promotes survivin-mediated

survival of TNBC cells. Cell Death and Disease. 2018;**11**:1-4. DOI: 10.1038/s41419-018-1083-y

[68] Han XJ, Yang ZJ, Jiang LP, Wei YF, Liao MF, Qian Y, et al. Mitochondrial dynamics regulates hypoxia-induced migration and antineoplastic activity of cisplatin in breast cancer cells. International Journal of Oncology. 2015;**46**:691-700. DOI: 10.3892/ijo.2014.2781

[69] Chen Z, Ai L, Mboge MY, Tu C, McKenna R, Brown KD, et al. Differential expression and function of CAIX and CAXII in breast cancer: A comparison between tumorgraft models and cells. PLoS One. 2018;**13**:e0199476. DOI: 10.1371/journal.pone.0199476

[70] Xu H, Eirew P, Mullaly SC, Aparicio S. The omics of triple-negative breast cancers. Clinical Chemistry. 2014;**60**:122-133. DOI: 10.1373/clinchem.2013.207167

[71] Weiner-Gorzel K, Murphy M. Mitochondrial dynamics, a new therapeutic target for triple negative breast cancer. Biochimica et Biophysica Acta (BBA)-reviews on. Cancer. 2021;**1875**(2):188518. DOI: 10.1016/j.bbcan.2021.188518

[72] Leites EP, Morais VA. Mitochondrial quality control pathways: PINK1 acts as a gatekeeper. Biochemical and Biophysical Research Communications. 2018;**500**:45-50. DOI: 10.1016/j.bbrc.2017.06.096

[73] Ge Y, Shi X, Boopathy S, McDonald J, Smith AW, Chao LH. Two forms of Opa1 cooperate to complete fusion of the mitochondrial inner-membrane. eLife. 2020;**9**:e50973. DOI: 10.7554/eLife.50973

[74] Yu H, Sun C, Gong Q, Feng D. Mitochondria-associated endoplasmic reticulum membranes in breast cancer. Frontiers in Cell and Developmental Biology. 2021 Jan;**28**(9):629-669. DOI: 10.3389/fcell.2021.62966

[75] Friedman JR, Lackner LL, West M, Dibenedetto JR, Nunnari J, Voeltz GK. ER tubules mark sites of mitochondrial division. Science. 2011;**334**:358-362. DOI: 10.1126/science.1207385

[76] Tanaka T, Kutomi G, Kajiwara T, Kukita K, Kochin V, Kanaseki T. Cancer-associated oxidoreductase ERO1-alpha promotes immune escape through up-regulation of PD-L1 in human breast cancer. Oncotarget. 2017;**8**:24706-24718. DOI: 10.18632/oncotarget.14960